MICRO TRIPS

894 EASY ADVENTURES FROM THE WORLD'S FAVORITE CITIES

HISTORY

FESTIVALS & EVENTS

FOOD & DRINK

CONTENTS

OUTDOORS

MUSIC & FILM

ARTS & CULTURE

INTRODUCTION

We scrambled up the coastal path from Seaford Bay, grasping at rocks and tufts of grass as the bank got steeper. Climbing further, we reached the top of the headland and the cold sea air blasted through our lungs and our heads. When we rounded the next turn we could see them – the Seven Sisters – an undulating series of chalk cliffs on the coast of southeast England and our destination for the day. They were so white they seemed to illuminate our faces with their reflected glow. We opened a flask of coffee in celebration – it was still only 10am and we were pleased with ourselves.

On waking in our London flat that Saturday morning, the working week behind us but thoughts of to-do lists and emails still buzzing through our minds, we'd made the decision we needed out of the city. We needed an adventure, new scenery and fresh air to make their mark on our weekend. And here we were, only two and a half hours later, and we had the whole day ahead of us.

It's easy to get caught in the urban jungle; 'when a man is tired of London, he is tired of life,' Samuel Johnson famously (and somewhat irritatingly) quipped, and it's true that most of the 60 cities represented in this book could fill a lifetime with new experiences. But what Johnson should have added is that the city is only the starting point. Easily accessible by train, bus or car, there is a world beyond to explore: prehistoric sites of standing stones, stately cathedral towns, forests, vineyards, arts festivals, rural pubs and gorgeous cliff walks. All are just a short trip away.

In this book we hope to inspire you to look beyond the city limits for your next adventure. Whether you're in Chicago, Vancouver, Brisbane or Rome and whether you live there, work there, are on vacation or are simply passing through with a day to kill, we encourage you to widen your net. Just an hour and 40 minutes from Cape Town you can spy breaching whales from the cliff path at Hermanus (p9); within two hours of Manhattan you can be surfing at Rockaway Beach (p207); and if you find yourself in Běijīng with time on your hands, you can choose between rafting a scenic gorge, visiting Jin-era temples, or hiking along the Great Wall (pp34-35) – all are easy micro trips from the city.

Each of the 60 global cities in this book is presented with a map of the surrounding area, pinpointed with up to 18 of the most exciting, in-the-know things to do within three hours. These pinpoints are colour-coded by theme so you can easily find what you're interested in, be that outdoor pursuits, arts and culture, history, festivals and events, film and music, or food and drink. The corresponding entries are ordered by the time it takes to get there from the city centre, so whether you've got just a couple of hours or a whole weekend, you can find an adventure, if not on your doorstep, then just a micro trip from it.

MIDDLE
EAST -&-
AFRICA

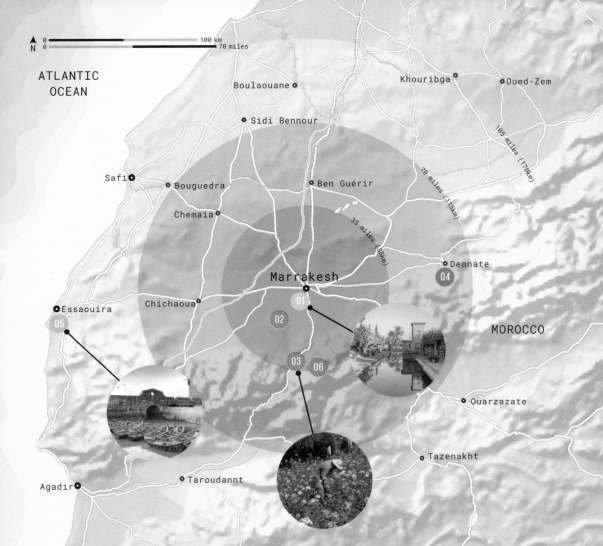

ATLANTIC
OCEAN

N
0 — 100 km
0 — 70 miles

Boulaouane

Khouribga ● Oued-Zem

Sidi Bennour

Safi ● Bouguedra ● Ben Guérir

Chemaia

105 miles (170km)

70 miles (115km)

35 miles (55km)

Demnate
04

Marrakesh

MOROCCO

01

02

Chichaoua

Essaouira
05

03 **06**

Ouarzazate

Tazenakht

Agadir ● Taroudannt

If you need a rest from Marrakesh's medina hubbub, it's easy to reach the snow-capped Atlas Mountains, where small Berber towns reveal a slower pace to Morocco. Short trips also lead to the Atlantic coast and even a mini-desert.

MARRAKESH

● ARTS & CULTURE   ● HISTORY   ● OUTDOORS   ● FOOD & DRINK   ● FESTIVALS & EVENTS   ● MUSIC & FILM

——— ONE HOUR FROM ———

01 Beldi Country Club

This chic retreat from the urban chaos offers a spa, a hammam, swimming pools, a bijou souq, and children's activities, from throwing pottery to baking bread, on a 14-hectare estate 6km south of town. Learn how to make a tagine the Moroccan way, or simply enjoy this green oasis of olive trees, rose gardens and Atlas views while lunching on Mediterranean and Maghrebi cuisine by the 35m pool. *www.beldicountryclub.com; 30min by petit taxi.*

02 Agafay Desert

Rather than braving a white-knuckle mountain pass to the Sahara, head 40km southwest of town for a mini-desert experience in Agafay's lunar landscape. Unlike in the Sahara, there's an artificial reservoir perfect for canoeing and, in spring, a carpet of wheat and wildflowers. Marrakshis come for weekends of horse-riding and dune-gazing at Agafay's eco-camps. *1hr by accommodation shuttle or hire car.*

——— TWO HOURS FROM ———

03 Ouirgane

Marrakshis in the know sneak 60km south to this bucolic Berber country retreat, fronting a lake en route to the Tizi n'Test mountain pass, for an accessible piece of the High Atlas. Drinking mint tea and gazing at the mountains await, as does hiking through unspoilt villages, and there are some chic places to stay and to eat. *2hr by grand taxi from outside Bab er-Rob.*

04 Demnate

Quite apart from its potteries using the distinctive local ochre clay, Demnate is an Atlas market town with a difference. Its interfaith heritage can be seen in the central *mellah* (Jewish quarter), two *zawiyas* (Islamic religious shrines) and Jewish and Islamic *moussems* (festivals), held in July and September respectively. It's also a great place to try local olives and honey. *2hr by grand taxi from the dirt car park outside Bab Doukkala.*

——— THREE HOURS FROM ———

05 Essaouira

Essaouira's wave-beaten ramparts and whitewashed medina featured in both *Game of Thrones* and Orson Welles' *Othello*. And adding to the port city's pop-cultural influence, the fort subsiding into the beach is said to have inspired Jimi Hendrix's 'Castles Made of Sand', and the guitarist apparently stayed in every riad in the medina. Eat grilled fish at the port and learn to kitesurf. *3hr by bus from CTM or Supratours station.*

06 Imlil

This Berber village, situated at an altitude of 1740m in the High Atlas foothills, is the main trailhead for treks up North Africa's highest peak, Jebel Toubkal (4167m). You can hike to the summit and back in two days, or climb just 200m past a burbling stream and orchards to the quiet village of Aroumd. Mountain-bike adventures are on offer, as are some excellent places to stay. *3hr by grand taxi from outside Bab er-Rob and minibus from Asni.*

TAXIS & SOUQS

An easy and affordable way to reach remote country towns is to time your visit to coincide with their weekly market days. The roads may be busy, which can be hairy if the town is up several narrow mountain passes, but you will have the best chance of finding a shared grand taxi, thanks to all the farmers and traders travelling to the souq. Tahanaoute and Asni's souq days fall on Tuesday and Saturday respectively, increasing grand taxi traffic in the Imlil area. Demnate's major Sunday souq is an event in itself, selling local products from olive oils and almonds to woodwork and henna-painted pottery.

BRILLIANT AFRICA

At the weekend, you'll find Africa on the beach – join the urban exodus in cars and

Muizenberg, South Africa

Sitting pretty on the well-trodden route from Table Mountain to Cape Point, this re-generated beach town offers multicoloured Victorian bathing huts, surf breaks (Agatha Christie learnt to surf here) and ice cream.
1hr from Cape Town by car.

Ponta do Ouro, Mozambique

Join the South Africans who cross the nearby border to this tropical getaway with a long, wide and surf-pound-ed beach. Don't miss diving or snorkelling to see dolphins, sharks and whales.
3hr from Maputo by car.

Serekunda, the Gambia

This dusty urban centre is the gateway to a string of Atlantic beach resorts, where a suntanned mix of expats, aid workers and British tourists hit the sand and the dance floor.
1hr from Banjul by car.

Kokrobite, Ghana

This beach getaway is a favourite of backpackers and volunteers for its long stretch of white sand, relaxed accommoda-tion and surf school. You can visit, while on a Ghanaian drumming and dance tour.
1hr from Accra by car.

El-Jadida, Morocco

The Portuguese must have had atmospher-ic weekend breaks in mind when they founded this 16th-cen-tury trading post, with its Unesco-protected old town, beaches and eerie vaulted cistern featured in Orson Welles' *Othello*.
1hr from Casablanca by car.

N BEACH TOWNS

shared taxis to the Atlantic, Indian, Med or Red.

Senga Bay, Malawi
Reggae drifts down the beach and fishing nets dry on the sand in this authentic spot on Lake Malawi. There's a winning hostel and a tropical fish farm, which breeds the lake's famous cichlids.
2hr from Lilongwe by car.

Grand Popo, Benin
The birthplace of voodoo may not sound like an ideal place to decompress, but it comes with the territory in Benin, and this former slave port abutting Togo has sultry palm-fringed beach ambience nailed.
2hr from Cotonou by car.

Bizerte, Tunisia
Head up the Mediterranean coast to Africa's northernmost city, which is also Tunisia's oldest and most European city, with a picturesque old port and medina, beautiful beaches, an atmospheric corniche and great seafood restaurants.
1hr from Tunis by car.

Stone Town, Tanzania
Hop on a ferry from Dar to the 'spice island' of Zanzibar, where glorious white-sand beaches surround Stone Town's warren of winding streets, markets, Swahili culture and Arab-built 17th-century fort.
2hr from Dar es Salaam by car.

Île de Gorée, Senegal
Once a slave-trading centre, Unesco-listed Gorée Island is today a slow-paced escape from the Senegalese capital, where locals sell craftwork between the bougainvillea-draped colonial edifices, and kids splash in the harbour.
1hr from Dakar by car.

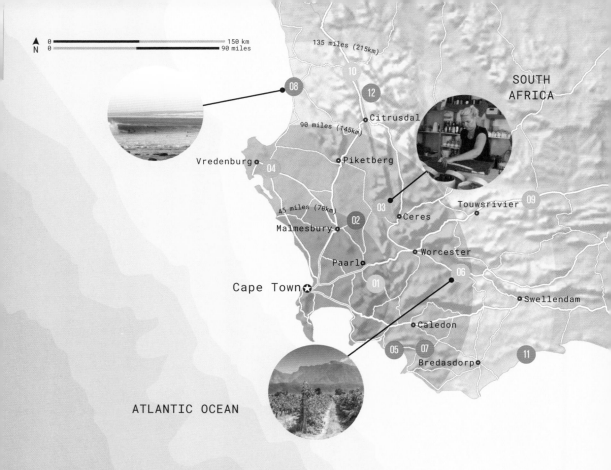

SOUTH
AFRICA

08

10

12

Citrusdal

90 miles (145km)

Vredenburg

04

Piketberg

03

Touwsrivier

09

Ceres

45 miles (70km)

02

Worcester

Malmesbury

06

Paarl

Swellendam

Cape Town

01

Caledon

05

07

Bredasdorp

11

ATLANTIC OCEAN

N

0   150 km
0   90 miles

Tear yourself away from Cape Town and you'll find testing surf spots, hikes along sand dunes, coastal paths and mountain ridges, charming historical hamlets and enough beer and wine tasting to replenish you after a long day.

CAPE TOWN

⬤ ARTS & CULTURE ⬤ HISTORY ⬤ OUTDOORS ⬤ FOOD & DRINK ⬤ FESTIVALS & EVENTS ⬤ MUSIC & FILM

———— ONE HOUR FROM ————

01 **Franschhoek Beer Company**
Franschhoek's economy is built on gastronomy and wine, but it is slowly becoming a beery destination, thanks to four excellent breweries in the area. Franschhoek Beer Company offers the perfect package – a great array of beers, innovative bites to pair with them, a stylish taproom and, for those travelling with the brood, a shaded kids' play area under ancient trees. *www.franschhoekbeerco.co.za; 10am–6pm Tue–Sun; 1hr 10min by car.*

02 **Riebeek Kasteel**
It's a rare visitor who can't find something to fall in love with in Riebeek Kasteel. Hidden courtyards harbour buzzing cafes where you can taste produce sourced from the local coffee roastery, olive farms, wine estates and chocolate shops. Once your belly is full, browse the galleries and pick up some ceramics, paintings, designer clothing or carpentry crafted by artists who find inspiration in this compact, pretty town. *www.riebeekvalley.info; 1hr 20min by car.*

03 **Church St, Tulbagh**
Tulbagh's Church St is known for its national monuments, but you don't have to be an architecture buff to appreciate its charms. Information panels transform Church St into an open-air museum, filling you in on local history as you wander between Victorian homesteads, celebrated country restaurants, shops, Cape Dutch hotels and a museum documenting the 1969 earthquake that almost destroyed it all. *www.tulbaghtourism.co.za; 1hr 20min by car.*

04 **West Coast Fossil Park, Langebaan**
Most people visit the West Coast for its white-sand beaches and seafood restaurants, or to windsurf on the lagoon. Veer inland instead for this passionately run fossil park, where remains of short-necked giraffes and sabre-toothed cats have been excavated. It's thought to be one of the richest fossil sites in the world, but, strangely enough, even the majority of locals have no idea it exists. *www.fossilpark.org.za; 8am–4pm Mon–Fri, 10am–1pm Sat; 1hr 30min by car.*

———— TWO HOURS FROM ————

05 **Cliff Path, Hermanus**
Hermanus is a particularly popular destination, and in whale-watching season the old harbour can get crowded with holidaymakers keeping an eye out for whales from the comfort of a seaside restaurant. But however many times you've visited, and however busy it is, you'll find beauty and solace on the 12km path that dips and winds along the town's diminutive cliffs. *www.hermanustourism.info; 1hr 40min by car.*

06 **Wacky Wine Weekend, Robertson**
The Robertson Wine Valley comes together to host this annual homage to the grape. Visit estates in Robertson, Ashton, Bonnievale and McGregor for tutored tastings, blending competitions, vineyard tours and unpretentious food-pairing experiences. Grab a weekend passport and hop between estates for live music and kid-friendly fun as well as plenty of opportunities to sip. *www.wackywineweekend.com; early Jun; 2hr by bus or car.*

FLOWER POWER

The Northern Cape region of Namakwa is the most famous area for those hoping to photograph the annual wildflower spectacular, but there are plenty of places to petal-watch closer to Cape Town. Darling (1hr by car) is known for its springtime wildflower show, while the West Coast National Park (1hr 30min by car) has blooms to rival those of the Northern Cape. A little further north, Clanwilliam (2hr 40min by car) is also a prime place to witness a normally barren landscape burst into colour. The best time to seek out the flowers is from early August to mid-September.

Klein River, Stanford

The Klein River is the star attraction of sleepy Stanford, which is a favourite destination for Cape Town weekenders. You can picnic on its banks, stroll the riverside footpaths or rent kayaks, but probably the best way to get acquainted with the river is on an afternoon boat trip. The skipper lights the on-board braai (barbecue) while pointing out the resident birdlife as you chug along. *www.stanfordinfo.co.za; 2hr by car.*

08 Elands Bay

This ramshackle little town is barely a town at all – more a selection of holiday homes, a fairly basic hotel and a couple of cafes and shops that seem to open whenever the owners aren't out surfing. Elands Bay is not the place to come if you're looking for luxury. This is a rustic escape for fans of windswept beaches, challenging surf and DIY entertainment at the end of the day. *2hr 30min by car.*

——— THREE HOURS FROM ———

09 Matjiesfontein

Head to the weird and wonderful Matjiesfontein for a glimpse into the past. The Victorian railway siding, founded in 1884 by Scottish railway porter James Douglas Logan, has the air of a slightly decrepit theme park, which is exactly its appeal. Staff dressed in period costume serve hearty slabs of cake in the cafe, tickle the ivories in the historical pub and show you around the eerie museum and its impressive collection of old tat. *www.matjiesfontein.com; 2hr 40min by car.*

10 Rooibos Teahouse, Clanwilliam

Put simply, rooibos tea is a South African symbol. The plant grows only in a tiny portion of the Cederberg region, and two local women have embraced their national treasure with this charming teahouse, serving more than 100 versions of the endemic infusion. Continue your education on the Rooibos Route, which kicks off at the teahouse. *www.rooibosteahouse.co.za; 8am–5pm Mon–Fri, to 2pm Sat; 2hr 40min by car or 3hr by bus.*

© Blaine Harrington / age fotostock / Alamy Stock Photo

De Hoop Nature Reserve
Perched on the south coast at the end of a dusty dirt road, De Hoop lies off the typically trodden path and is all the more magnificent for it. It's tough to choose the reserve's finest feature: the untouched, dune-backed beaches; the winter whale sightings; glimpses of the endangered Cape vulture soaring overhead; or wildlife walks with the chance of spotting zebras, baboons and bonteboks. *www.capenature.co.za; 7am–6pm; 3hr by car.*

 Algeria Forest Station, Cederberg
In a little more than three hours' drive from the city you can be high up in the Cederberg Mountains, where the weirdly shaped sandstone rock formations are home to some first-rate hiking trails and climbing spots. There are caves to explore, ancient rock paintings to view and, after a long and dusty hike in the heat, a superb winery to visit. *www.capenature.co.za; 3hr 10min by car.*

'Friends are constantly asking when they can join me for a weekend in Tulbagh. Our family has a small holiday house there and I tend to rave about the place. There's something about the fresh country air that helps me disconnect from busy city life and sleep like a champ. My ideal weekend would probably be visiting a wine farm in the area; my two favourites are without a doubt Twee Jonge Gezellen and Montpellier. When I'm not feeling lazy I sometimes tackle a mountain-biking trail through the vineyards to energise before heading back to my home in the city centre.'

Troye May, radio host

11

© Ariadne Van Zandbergen / Alamy Stock Photo

EASY-TO-REACH AFRICAN WILD

The word 'safari' originates from the Arabic *safara* ('to travel'). While many of Africa's best known safari spots demand a lengthy trip, there are plenty of wildlife-watching experiences within reach of key cities.

LIFE WATCHING

Pilanesberg National Park, South Africa

Switch off the engine and listen to white rhinos munching just metres away, while elephant herds plod along in the distance and lions lurk in the undergrowth. *www.pilanesberg nationalpark.org; 6am-6pm Mar-Oct, 5.30am-7pm Nov-Feb; adult/child R110/30; 3hr 20min from Johannesburg.*

Abuko Nature Reserve, the Gambia

Sightings here are modest compared to larger parks, but the chance to walk around the reserve makes it a special experience. Keep an eye out for crocodiles, snakes and a range of monkeys. *8am-5.30pm; adult/child US$0.70/0.30; 1hr from Banjul by car.*

Crescent Island, Kenya

There are no predators on Crescent Island, meaning you can walk freely among the giraffes and zebras. Boats convey visitors from the shores of Lake Naivasha; look out for hippos en route. *Naivasha Bottom Rd, Naivasha; 8am-6pm; US$30; 2hr 10min from Nairobi by car.*

Volcanoes National Park, Rwanda

It's difficult to put into words the thrilling experience of scrambling through the jungle to find mountain gorillas grooming, playing and observing you, with almost as much curiosity as you have for them. *www. volcanoesnational- park.org; gorilla trek US$1500; 2hr 30min from Kigali by car.*

Matobo National Park, Zimbabwe

Wildlife, nature and history meet in Matobo. The majestic balancing rocks are the backdrop for superb rhino sightings, while ancient San rock art offers insight into the Matobo of yore. *www.zimparks.org; 6am-6pm; US$15; 1hr from Bulawayo by car.*

Hluhluwe-iMfolozi National Park, South Africa

The rolling hills of this scenic park offer excellent vantage points for spotting giraffes, buffalo, elephants and rhinos, both black and white. Book way ahead for an unforgettable multi-day guided trek. *www.kznwildlife.com; 5am-7pm Nov-Feb, 6am-6pm Mar-Oct; adult/child R220/110; 2hr 40min from Durban.*

Tacugama Chimpanzee Sanctuary, Sierra Leone

Although the animals here are not strictly roaming wild, the 90-minute tour of the sanctuary is a must for anyone visiting Freetown. Extend your stay in one of the rustic lodges, each named in memory of a chimp. *www.tacugama.com; tours 10.30am & 4pm; adult/child US$15/5; 30min from Freetown by car.*

Watamu Marine National Park, Kenya

Reached by glass-bottomed boat, Watamu offers wildlife-viewing of the sub-aquatic kind. Don a snorkel and glimpse turtles and tropical fish darting between coral reefs. *www.kws.go.ke; 8am-6pm; adult/child US$17/13; 2hr from Mombasa by car.*

Mokolodi National Reserve, Botswana

The highlight of this small reserve is the chance to track rhino and giraffe on foot on a half-day guided adventure. Alternatively, you can cycle or join an organised drive. *www.mokolodi.com; 7.30am-6pm; rhino tracking P590; 40min from Gaborone by car.*

Lower Zambezi National Park, Zambia

Ditch the wheels in favour of a canoe safari along the Zambezi River. There's nothing quite like spotting hippo, elephant or even a lion pride as you glide along the water. *www.lowerzambezi. com; 6am-6pm Apr-Nov; US$25; 2hr from Lusaka by car.*

DUBAI

From the tallest mountain in the UAE to the dusty desert floor, from the crystal-clear sea to archaeological discoveries, you'll find huge contrasts to the glitz of Dubai just outside the city.

Map labels:

N

0 — 200 km
0 — 120 miles

IRAN

PERSIAN GULF
(ARABIAN GULF)

Bandar Abbas

15
13
07
14
08
09

Ras Al Khaimah
Umm Al Quwain
01 03
Dubai 02
04
Jebel Ali

Fujairah City

GULF OF OMAN

06
05
ABU DHABI

60 miles (95km)
Al Ain
10 11
12

Suhar

As Sib

UNITED ARAB EMIRATES

120 miles (195km)

16

Ibri

17
18

180 miles (290km)

OMAN

SAUDI ARABIA

● ARTS & CULTURE   ● HISTORY   ● OUTDOORS   ● FOOD & DRINK   ● FESTIVALS & EVENTS   ● MUSIC & FILM

——— ONE HOUR FROM ———

01   **Sharjah Art Scene**
Keep tabs on what's hot in Middle Eastern contemporary art by checking out the collection of the renowned Sharjah Art Museum, followed by a spin around the cluster of modern white-cube galleries in the Heart of Sharjah heritage area. The latter is part of the Sharjah Art Foundation, organiser of the prestigious Sharjah Biennial since 1993. *www.sharjahart.org; 30min by car or up to 1hr by bus.*

02   **Wasit Wetland Centre**
Incredibly enough, this compact bird haven in Sharjah, about 35km north of Dubai, was reclaimed from an illegal dumping ground. Today it's a fabulous spot to learn about and observe dozens of feathered friends, including flamingoes, herons, ducks and less familiar species such as the endangered northern bald ibis. The birds make their homes both in aviaries and among the dunes, salt flats and lagoons of the wetlands area. *Wed–Mon; 40min by car.*

03   **Al Zorah Nature Reserve**
The quiet is punctuated only by squawking seabirds as you paddle softly around the shallow waters of this tidal lagoon and mangrove forest 50km north of Dubai. The natural highlight of Al Zorah, a new community taking shape along 12km of waterfront, this unspoilt ecosystem hosts pink flamingoes, egrets, herons and dozens of other bird species, and also doubles as a fish nursery. *www.alzorah.ae; 50min by car.*

04   **Mleiha Archaeological Centre**
The sun-baked Sharjah sands have yielded some of the most remarkable artefacts to be unearthed in the UAE, including Paleolithic stone tools, which suggest that early Homo sapiens crossed the region en route from Africa. A modern exhibition centre showcases the finest finds and also serves as a launch pad for tours to Stone-Age caves, ancient burial sites and stunning rock formations. *www.discovermleiha.ae; Sat–Wed; 1hr by car.*

05   **Abu Dhabi Pearl Journey**
Pearls may be precious, but the life of pearl divers was hard graft, as you'll learn on a tour aboard a *jalboot* (traditional boat), meandering around the mangroves in northern Abu Dhabi, some 130km south of Dubai. Scan the skies for herons, egrets and other seabirds, before opening your own oysters. If you happen upon a pearl – it's finders keepers. *www.adpearljourney.com; 9am–7pm; 1hr 20min by car.*

06   **Nurai Island**
Bling meets barefoot on this royally owned island a mere 10-minute boat ride from Abu Dhabi. If you can't stretch to a villa that has a private pool, score a day pass to luxuriate on the pearly white beach, drinking in blissful azure sea views, cool cocktail in hand. After a sunset meal return to land – and reality – under a canopy of twinkling stars. *www.zayanuraiisland.com; 1hr 30min by car.*

THE LOCAL'S VIEW

'My favourite Dubai getaway is one that speeds through the spine of the desert and, in a short 45 minutes, takes you to the sleepy town of Ras Al Khaimah. Here the skyline is clear and you can take part in simple pleasures, such as a boat ride in the mangroves; a walk through the date plantation at Khatt, ending with tasting the freshest produce; and a sit-down (on the floor) traditional Yemeni *mandi* meal with the locals (at Mahareb Mandi, for instance).'

Abhiroop Sen, director of communications strategy

 12

07 Dhayah Fort

A sweeping panorama of palm plantations, mountain peaks, city lights and the placid Gulf unfolds below a rocky outcrop crowned by this early 19th-century fortress in the northern emirate of Ras Al Khaimah. Although it ultimately failed to fend off British invaders, the historic structure is well worth the climb up a zigzagging stone staircase, especially at sunset. *1hr 30min by car.*

08 Hatta Hiking Trail

The Dubai mountain exclave of Hatta has been busy sharpening its outdoorsy profile. A 9km hiking trail, opened in 2017, meanders from town to the tranquil emerald waters of the Hatta Dam, where you can punt around by kayak or continue into the hills for another 3km. The route skirts the Hatta Heritage Village, a farmers' market and a park. *1hr 30min by car or 3hr by bus from Sabkha station.*

09 Hatta Mountain-Bike Trail

Strap on your helmet and two-wheel it through spectacularly stark scenery at the foot of the Hajar Mountains, some 135km southeast of Dubai. The bike park, which is part of an effort to boost Hatta's outdoor pursuits potential, consists of 50km of well-marked single-track trails, colour-coded by level of difficulty. On weekends, bikes can be rented at the trail centre. *www.hattamtb.ae; 1hr 30min by car.*

--- TWO HOURS FROM ---

10 Al Ain Oasis

Slap bang in the heart of Al Ain, about 150km southeast of Dubai, this forest of 147,000 date palm trees welcomes visitors with soothing shade and abundant greenery. Wells and an indigenous irrigation system called *falaj* feed this massive oasis, which became the UAE's first Unesco World Heritage Site in 2011. Learn more at the interactive Eco-Centre and from the interpretive panels lining the walkways. *www.visitabudhabi.ae; 8am–5pm; 1h 40min by car.*

11 Al Ain Palace Museum

Draw back the curtain on the private lives of Bedouin royalty at this rambling 1937 fort where the UAE's founding father, Sheikh Zayed, lived, loved and ruled in the pre-oil era. Snap a selfie with the pair of wedding-cake towers flanking the entrance, then wander into bedrooms, the kitchen, guest quarters and other rooms, most of them beautifully preserved. *www.visitabudhabi.ae; Tue–Sun 8.30am–7.30pm, from 3.30pm Fri; 1h 40min by car.*

12 Jebel Hafeet

Looming above Al Ain, 1240m-high Jebel Hafeet is not only the UAE's second-highest mountain but also the spiritual guardian of 500 Bronze-Age tombs tucked into its foothills. The 12km tarmac corkscrewing to the top makes for one of the country's most spectacular drives. Scan the craggy terrain for red foxes, feral cats and rabbit-like rock hyraxes which are, improbably, related to the elephant. *2hr by car.*

13 Jebel Al Jais Mountain Road

The 30km drive up Jebel Al Jais, at 1934m the highest mountain in the UAE, is one of the country's most mesmerising road trips. Snaking up from the desert floor, this smooth highway to heaven treats you to spellbinding views of fissured mountain faces and deep canyons after every bend. Lay-by areas along the way and a viewing deck near the summit invite picnics or quiet contemplation. *www.jebeljais.ae; 2hr by car.*

14

14 Al Aqah Beach

Count the clownfish, track elusive sea turtles or spot reef sharks as you dart around the reefs, wrecks and rich waters surrounding Snoopy Island off Al Aqah, in the eastern emirate of Fujairah. In this unhurried and remote part of the country, the mighty Hajar Mountains cascade down to talcum beaches lapped by the cobalt-blue Indian Ocean. En route, stop at Al Badiyah, the UAE's oldest mosque. *2hr by car.*

15 Musandam Peninsula

The Omani exclave of Musandam has been dubbed the 'Norway of Arabia' for its fjord-like inlets hugged by sheer honey-coloured cliffs. Admire this Martian landscape from a wooden *dhow* (traditional boat), keeping an eye out for dolphins and cormorants and dipping into the aquamarine sea for a swim or a snorkel. Boats set sail from Khasab but it's easiest to book a tour package, including transport from Dubai. *2hr 30min by car.*

16 Al Dhafra Festival

Every December, long legs, shapely humps and thick eyelashes turn heads in ho-hum Madinat Zayed, 260km southwest of Dubai. The camel beauty pageant is the flagship event of this 11-day celebration of Emirati heritage, along with camel races, a traditional souq, and competitions involving falcons, classic cars, Arabian horses and *salukis* (Arabian dogs). *www.turathuna. ae/en/event/al-dhafra-festival; Dec; 2hr 30min by car.*

——— THREE HOURS FROM ———

 Liwa Oasis

Vast open spaces lidded by cornflower skies. Undulating dunes shimmering in shades from silver to cinnamon. Lone camels by the roadside and lush date palm groves. Expect all this and more on an overnight trip (consider camping) to the Liwa Oasis, the ancestral land of the ruling families of Dubai and Abu Dhabi. It's on the edge of the Empty Quarter desert, 340km south of Dubai. *3hr 30min by car.*

18 **Moreeb Dune**

Almost nipping at Saudi territory, Tel Moreeb soars nearly 300m, making it one of the world's tallest sand dunes. It's an eerily quiet spot, except during February's Liwa International Festival when supercharged SUVs race each other up this near-vertical wall of sand. Prepare to sweat buckets as you trek to the top, ideally at sunset, where you'll be rewarded with panoramic views and a mystical interplay of light and shadow. *3hr 30min by car.*

CAMEL RACES

Camel racing is rooted in the Emirati soul, and attending a meet is popular with locals and visitors alike. It's quite an exhilarating sight when hundreds of one-humped dromedaries dash out of their pens and on to the dirt track, jostling for position in a lumbering gallop with legs splayed in all directions. Racing season lasts from November to April at tracks throughout the UAE. A highlight is the Al Marmoom Heritage Festival, a four-week celebration held at Al Marmoom Village, 40km southeast of Dubai, which also features music, crafts, food stands and carnival rides. *www.almarmoom festivals.ae; 40min by car.*

18

© Horst Bottner / 500 px

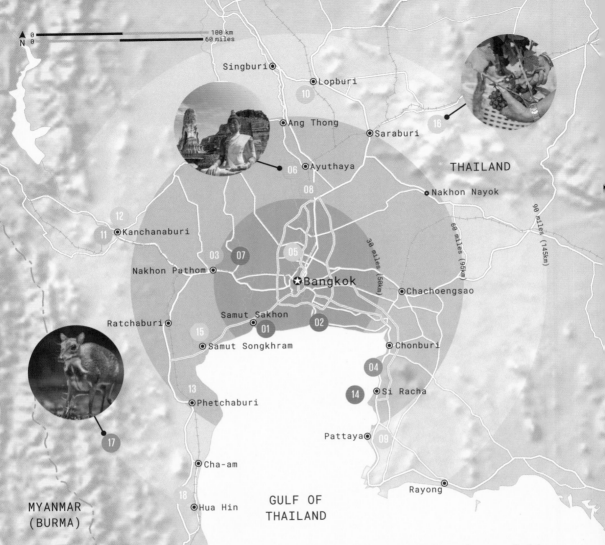

N

0
0
100 km
60 miles

Singburi

Lopburi

10

Ang Thong

Saraburi

06 Ayuthaya

THAILAND

08

Nakhon Nayok

16

12

11 Kanchanaburi

03 07

05

30 miles (50km)

60 miles (95km)

90 miles (145km)

Nakhon Pathom

★ Bangkok

Chachoengsao

Samut Sakhon

Ratchaburi

15

01

02

Chonburi

Samut Songkhram

04

13

14 Si Racha

Phetchaburi

Pattaya

09

17

Cha-am

Rayong

18

Hua Hin

GULF OF
THAILAND

MYANMAR
(BURMA)

A sprawl of concrete and canals, temples and palaces, atop a base layer of old-
fashioned wooden homes – Bangkok is big. Those escaping its pull are rewarded
with beaches, jungles, tropical islands and temples, all within striking distance.

BANGKOK

● ARTS & CULTURE　　● HISTORY　　● OUTDOORS　　● FOOD & DRINK　　● FESTIVALS & EVENTS　　● MUSIC & FILM

—— ONE HOUR FROM ——

01 Samut Sakhon

Slumbering Samut Sakhon is usually visited en route to Amphawa, but it's more than worth a trip in its own right in order to enjoy a soothing dose of Thai country life. Wooden fishing boats adorn the harbour, and, across the inlet, Wat Chawng Lom is dominated by a 10m-high statue of Guan Yin, the Chinese Goddess of Mercy – a nod to the days when junks flocked here laden with goods from the South China Sea. *1hr by train from Thonburi.*

02 Ancient City, Muang Boran

On paper, the Ancient City sounds slightly tacky – a scaled-down recreation of famous sites from across the nation, scattered across a Thailand-shaped water garden at the mouth of Chao Phraya River – but the lack of crowds and the fine craftwork make it rather charming. Hire a bike and pedal past mock-ups of Rattanakosin, Ayuthaya, Chiang Mai and Sukothai. *9am–7pm; 1hr by BTS (Skytrain) & shuttle bus.*

03 Phra Pathom Chedi, Nakhon Pathom

A proper stupa should be visible from miles around, soaring against the sky like a gilded exclamation mark, and so it is with Phra Pathom Chedi, which has a 127m spire first raised in the 6th century. It's since been a simple brick stupa, an ornate Khmer *prang* (tower) and a picturesque ruin, before re-emerging as a handsome bell-shaped chedi in 1860. *Daylight hours; 1hr by minivan from Southern Bus Terminal.*

04 Bang Saen Beach

As the closest beach to Bangkok, Bang Saen might lead you to expect Khao-San-Road-on-Sea, but in fact most overseas visitors head further afield to the southern beaches, leaving the place to the locals. You'll find Thai families floating on rubber rings by day, while students gather to sink pitchers and feast on cheap eats after dark. It's no Waikiki, but also no Pattaya, and that has to be a good thing. *1hr by bus from Eastern Bus Terminal.*

05 Chit Beer, Ko Kret

Due to Thailand's tight licensing regulations, brewing craft beer is a guerrilla enterprise in Bangkok. Tucked away in a village house on the potters' island of Ko Kret, Chit Beer produces full-bodied ales on a home-brew scale, with beers sold straight from the tap, views out over the canal and a rebel vibe. *www.facebook.com/chitbeer; by appt; 1hr 30min by bus from Victory Monument & boat from Wat Sanam Neua.*

06 Wat Mahathat, Ayuthaya

Bangkok has more ancient wat than you can shake an incense stick at, but most of these have been restored so many times over the centuries that they gleam like gemstones. Not so Wat Mahathat, the most dramatic of the Khmer-influenced temples dotted around Ayuthaya. Founded in 1374, its central *prang*, which collapsed for a second and conclusive time in 1911, was once 43m high. Royal ghosts drift among the ruined plinths and the face of Buddha stares out from a tangled web of tree roots. *8am–6.30pm; 1hr 30min by train from Hualamphong.*

VENICE WITH NOODLES?

It isn't immediately obvious today, but Bangkok was built on a network of canals, linking the royal island of Rattanakosin to the inlets flowing out of the surrounding paddy fields into the Gulf of Thailand. Up until the 19th century, boats were the main form of transport in this mighty metropolis, and visiting seafarers dubbed Bangkok 'the Venice of the East'. Today, most of the canals snaking through the centre have been filled in for road-building, but you can still charter a longtail river taxi to explore the *klorngs* (channels) that weave like a spiderweb through the suburbs.

07 Wat Bang Phra, Nakhon Chaisi
A short hop from the city limits, but far from the cacophony of Banglamphu, this provincial wat in sleepy Nakhon Chaisi is one of the country's leading centres for *sak yant* tattooing. Devotees and a scattering of nervous tourists queue up daily for intricate spiritual designs, etched the painful way with a tapped bamboo needle. *Daylight hours; 2hr by bus & sŏrng·tăa·ou from Northern & Northeastern Bus Terminal.*

08 Bang Pa In Palace
What the builders of Bang Pa In lacked in subtlety, they made up for in ambition. Originally founded in the 17th century, the palace was expanded and ornamented with European-style baroque trim by kings Rama IV and V. Today, in between the Thai monastery buildings and eccentric, Chinese-influenced towers, you'll find ice-white wedding cake pavilions and statues of Renaissance women. *8am–4pm; 2hr by train from Hualamphong.*

09 Mantra, Pattaya
Not everyone comes to Pattaya for the go-go bars. Plenty of Bangkok residents buzz down for the beaches and the food – and at Mantra, the food on offer is extremely good. Seven open kitchens serve up everything from salmon sashimi to roast duck red curry in a New York loft-style interior, which is adorned with wooden fretwork screens and alcoves full of trinkets. *www.mantra-pattaya.com; 5pm–1am, plus 11.30am–3pm Sun; 2hr by bus from Eastern Bus Terminal.*

© Infinity T29 / Shutterstock

10 Prang Sam Yot, Lopburi
It's easy to overdose on Thai wat and Khmer *prangs* (temple towers), but at Prang Sam Yot, the resident monkeys are the main attraction. In tribute to the temple's Hindu origins, rhesus macaques have free run of the compound, particularly during November, when the whole place goes monkey-crazy and a huge buffet is laid on for the temple's simian attendants. *8.30am–6pm; 2hr 30min by train from Hualamphong.*

11 Death Railway Bridge, Kanchanaburi
This pivotal crossing on the Siam–Burma Railway was built using the forced labour of thousands of civilians and prisoners of war. Considering the violence of its construction, the bridge is a remarkably calm vantage point from which to contemplate this dark chapter in Thai history. *24hr; 2hr 30min by bus from Mo Chit.*

12 On's Thai-Issan, Kanchanaburi

Most people come to Kanchanaburi for the history rather than the cooking, but there's good eating to be had in the lanes flanking the Khwae Yai River. On's Thai-Issan delivers a full spread of Isan dishes made from purely vegetarian ingredients – banana flowers, ginger tofu, brown rice and the like – served in generous portions in a pint-sized, hole-in-the-wall dining room. *http://onsthaiissan.com; noon-10pm; 2hr 30min by bus from Mo Chit.*

13 Phra Nakhon Khiri Historical Park, Phetchaburi

Never one for consistency, King Rama IV chose a curious medley of Thai, Chinese and European buildings for his Phetchaburi retreat, crowning a forested hill on the west side of town. Gratifyingly few tourists make it here, so enjoy the quiet as you stroll between the trees to three handsome stupas that lord it over the surrounding countryside. *8.30am–4.30pm; 2hr 30min by train from Hualamphong.*

FESTIVAL DETOURS

Thai festivals are spectacular, marked by riotous pageantry and fabulous food. Bangkok is typically the life and soul of the party but you can celebrate in style elsewhere. For a festival soaking in a stunning setting, celebrate Songkran, the Thai New Year in April, in ancient Ayuthaya (1hr 30min by train). October is the time of buffalo races in Chonburi (2hr by bus), while a lavish feast is laid on for the monkeys at Prang Sam Yot in Lopburi in November (2hr 30min by train). For something more modern, try Wonderfruit (*pictured left*), Thailand's biggest festival of pop music, food and the arts, which is held near Pattaya in December (2hr by bus).

© Courtesy of Wonderfruit

14 **Ko Si Chang, Si Racha**
Sense your pulse slow on the pleasure cruise from the fishing port of Si Racha to the one-time royal beach retreat of Ko Si Chang. Come during the week and the chaos of the mainland feels far behind as you paddle on the beach and wander up to the dragon-filled temple on the hill above the shore. *2hr 50min by bus from Eastern Bus Terminal and boat from Si Racha.*

15 **Tha Kha Floating Market, Amphawa**
Zip out to sleepy Samut Songkhram, pick up a *sŏrng·tăa·ou* to a backwater *klorng* (canal) and discover the least touristy floating market within reach of Bangkok. Time it right and you'll arrive to find real vendors selling real produce – lotus flowers, rambutans, bitter melons, fish steamed in banana leaves – to real customers, from a flotilla of canoes. *Market times vary; 2hr 50min by bus & sŏrng·tăa·ou from Southern Bus Terminal.*

16 **GranMonte Vineyard & Winery, Nakhon Ratchasima**
Thai wine has come a long way since the first vines were planted near Hua Hin in 2002 (by the makers of Red Bull). GranMonte is one of the new school of Thai vineyards, teasing quality, award-winning Chenin Blanc, Shiraz and Viognier from the dry hills around Khao Yai. It's a family affair and the crew includes Thailand's first female oenologist. *www.granmonte.com; tours 11am–3pm, 10am–4pm Sat & Sun; 2hr 30min by car.*

© Komut / Shutterstock

© nimon / Shutterstock

Kaeng Krachan National Park
You don't have to travel the length of the peninsula to find true wilderness. Kaeng Krachan National Park has it all – elephants, tigers, leopards, gibbons, tapir and 400 species of birds, all hidden by 2915 sq km of lush rainforest. Finding the residents takes a good guide and a bit of luck, but the noise of Bangkok becomes a distant memory once you enter this green bower. *3hr by car.*

Cicada Market, Hua Hin
Thailand's original seaside resort lives on its beach reputation, but the best of Hua Hin is back from the sand. The illuminated walkways of Cicada Market throng with buskers, street-food vendors and hawkers selling familiar wares without the big-city hard sell. Stock up, then cruise the lanes and feast on Hua Hin seafood. *4–11pm Fri–Sun; 3hr 30min by train from Hualamphong.*

17

© Prasit Chansareekorn / Getty Images

THE LOCAL'S VIEW

'Ayuthaya is my favourite day-trip escape from the frenetic pace of Bangkok life, and it's ideal for travellers wanting to delve into the history of the kingdom of Siam. From 1350 until 1767 the city reigned over the nation, before it was sacked by the Burmese, and the ruins of its 400 temples and palaces are fascinating to explore. To save time, come by minivan, then hire a bicycle to explore. Must-sees are Wat Ratchaburana; Wat Mahathat, famed for its sandstone Buddha head tangled in the roots of a Bodhi tree; and the 12.5m-tall bronze Buddha in Wihan Phra Mongkhon Bophit.'

Nardia Plumridge, travel writer

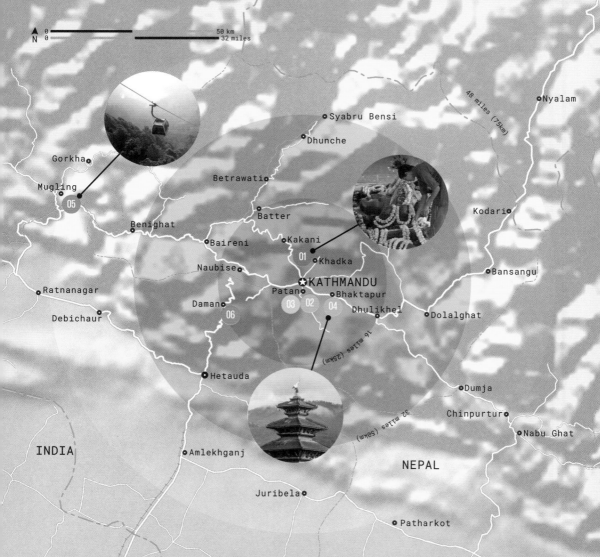

Scale:
- 0 — 50 km
- 0 — 32 miles

N

Nyalam

48 miles (75km)

Syabru Bensi

Dhunche

Gorkha

Mugling

05

Betrawati

Batter

Kakani

Kodari

Baireni

★ KATHMANDU

Khadka

Bansangu

Naubise

Patan

Bhaktapur

Benighat

01

03 02 04

Dhulikhel

Dolalghat

Ratnanagar

Daman

06

Debichaur

16 miles (25km)

Dumja

Hetauda

Chinpurtur

32 miles (50km)

Nabu Ghat

INDIA

Amlekhganj

NEPAL

Juribela

Patharkot

Hemmed in by the Himalaya, Kathmandu is cut off from the world. Escaping the city takes patience and a lot of twisting mountain roads. But you don't have to go far to find Shangri-La, among ancient townships, temples and mountain views.

KATHMANDU

● ARTS & CULTURE   ○ HISTORY   ○ OUTDOORS   ○ FOOD & DRINK   ○ FESTIVALS & EVENTS   ○ MUSIC & FILM

——— ONE HOUR FROM ———

01 Budhanilkantha
Devout Hindus believe that Vishnu resides through the ages on the coils of the serpent Ananta-Shesha, and this legend comes vividly to life at Budhanilkantha. Pilgrims gather at all hours to pay their respects at the largest monolithic statue in Nepal, a serene vision of deity and serpent executed in black stone, reclining at the bottom of a sacred tank. *1hr by bus from Ratna Park.*

02 Patan Museum
The past spills out of every courtyard in the Kathmandu Valley, so it's only appropriate that the country's finest historical treasures are displayed inside a piece of ancient history. Preserving an incredible collection of religious art and artefacts, the Patan Museum fills a wing of the sprawling palace of the Malla kings, a medieval fantasy of carved balconies, fretwork screens and deity-covered timbers. *www.patanmuseum.gov.np; 8am–6pm; 1hr on foot from Thamel.*

03 Newari Kitchen, Patan
Nine times out of 10, when you find Newari food in Kathmandu, it comes with a tacky dance show for tourists. Not so at Newari Kitchen. Here you'll find real Newari food being enjoyed by real Newari people, away from the tourist bustle in Pulchowk. Play it safe with gateway dishes such as *chatamari* (rice pancakes) or go all out for *shapo mhicha* – fried, tripe-wrapped bone marrow. *https://newarikitchen.business.site; noon–10pm; 1hr by rickshaw from Thamel.*

04 Nyatapola Temple, Bhaktapur
Towering temples are the symbol of Nepal, and the graceful Nyatapola Temple is the tallest of them all. Even the 2015 earthquake failed to diminish this medieval skyscraper, which rises in five graceful tiers above Taumadhi Tole. Etched with intricate carvings, this was once the highest building in the whole of Nepal, and it stands as a monument to Nepal's resilience in the face of disaster. *Dawn–dusk; 1hr by bus from Bagh Bazaar.*

——— THREE HOURS FROM ———

05 Manakamana Cable Car
Although the Manakamana temple has a 17th-century pedigree, it's not the architecture that steals the show, but the journey up here. A Swiss-engineered cable car – complete with a special carriage for sacrificial goats – climbs over 1000m from the Prithvi Hwy, offering awe-inspiring views over the Trisuli Valley. Rather than returning the same way, you can walk back to the valley floor in about three hours. *9am–5pm, from 8am Sat; 3hr by car.*

06 Daman View Tower, Daman
There may be closer viewpoints to Kathmandu, but for the full Himalayan panorama, there's really only one choice. Ignore the tower itself (a concrete throwback from the 1980s) and concentrate instead on the vista, with a curtain of 8000m peaks riding the horizon all the way from Dhaulagiri in the west to Everest in the east. Peaceful trails in the surrounding forest lead to hidden temples and Buddhist monasteries. *Daylight hours; 3hr by car.*

ONE VALLEY, THREE KINGDOMS

Kathmandu is the beating heart of Nepal but in medieval times the city faced a serious challenge from two rival city-states in the Kathmandu Valley. Just across the Bagmati River, the city of Patan was ruled by a competing clan, while a third dynasty controlled Bhaktapur to the east. From the 15th century, these close neighbours engaged in an epic game of 'keeping up with the Joneses', filling their cities with ever more elaborate temples and palaces – the origins of the grand Durbar Squares at the heart of Kathmandu, Patan and Bhaktapur today.

© Suman Acharya / Alamy Stock Photo; © Lindsay Brown / Alamy Stock Photo; © Oleskaus / Shutterstock

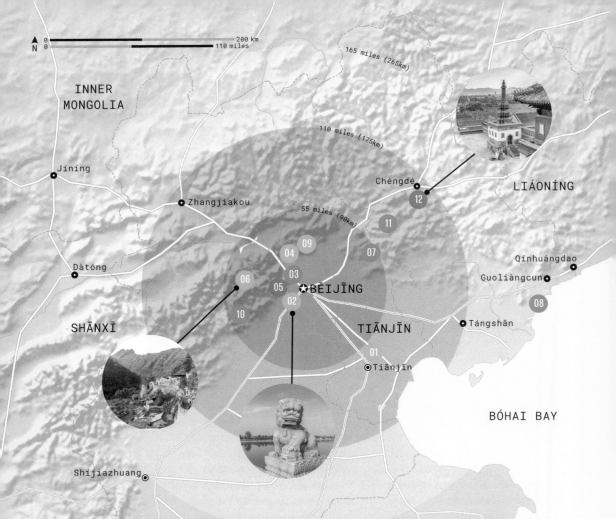

N

| 0 | 200 km |
| 0 | 110 miles |

INNER
MONGOLIA

165 miles (265km)

110 miles (175km)

Jíníng

Chéngdé

LIÁONÍNG

Zhangjiakou

55 miles (90km)

12

11

Qínhuángdao

Dàtóng

04   09

07

Guoliàngcun

06   03

★BĚIJĪNG

08

05

SHĀNXĪ

02

10

TIĀNJĪN

Tángshān

01

Tiānjǐn

BÓHAI BAY

Shijiazhuang

Escape Běijīng's urban sprawl via China's high-speed railway to discover valleys, mountains, tombs and pagodas. But it's not all about the Old World: you can explore a Zen library, hit the coast, relax in a spa or step out onto a glass bridge.

BĚIJĪNG

ARTS & CULTURE　　HISTORY　　OUTDOORS　　FOOD & DRINK　　FESTIVALS & EVENTS　　MUSIC & FILM

——— ONE HOUR FROM ———

01 Tiānjīn

The former foreign concession port of Tiānjīn is now a booming metropolis, just southeast of Běijīng, with laid-back European-style neighbourhoods, a riverside promenade and a thriving craft-beer scene. Head to the Jizhou district for two historic temples: the Dule Temple, dating to the Liao dynasty (907–1125) and Baita Temple, with its white pagoda believed to have been built in the 6th century. *30min by high-speed train from Běijīng South station.*

02 Marco Polo Bridge

This 900-year-old bridge in nearby Wǎnpíng Town is named after the great Italian traveller who described it in his diaries. Built in 1189 (and expanded in the mid-20th century), the bridge is adorned with 485 carved stone lions and spans the Yǒngdìng River, historically providing the main route into Běijīng from the southwest. *7am–6pm; 40min by subway line 14 to Dāwǎyáo station or by bus 662 from Chángchūnjiē station.*

03 Fragrant Hills Park

Within easy reach of Běijīng, the Western Hills were a popular retreat for many emperors. A number of temples and memorial halls dot the hills, notably the ancient but little-visited Azure Clouds Temple with its vast protector deity halls and bell and drum towers. Hike or take a chairlift to the top of Incense-Burner Peak for sweeping panoramas. *6am–6pm; 45min by subway to Xiyuan or Yuanmingyuan, then bus 331.*

04 Ming Tombs

The Unesco-listed site of 13 Ming dynasty emperors' tombs makes for a fascinating trip from Běijīng. Each tomb comprises an enormous temple-like complex with its own burial mound, all set on the slopes of Tiānshòu Mountain. Three of the tombs are open to the public and include the remains of not only the emperors but their concubines and an incredible set of artefacts. *8am–5.30pm; 1hr by Chángpíng line subway or bus 872 from Dengshemen Gateway.*

——— TWO HOURS FROM ———

05 Fahai Temple

Anyone remotely interested in Buddhist art should make the pilgrimage to this Ming-era temple, which houses some of the most striking but little-known Buddhist frescoes in China, which are comparable to those found at Yongle and Dūnhuáng. *9am–5pm; 1hr 30min–2hr by subway to Pínguǒyuán station, then a 10min bus and a 30min walk.*

06 Chuāndǐxià

A charming cluster of historic courtyard homes nestled in a valley, this Ming dynasty village is a welcome antidote to the scrum of the city some 90km away. Spend a few hours exploring the Old-World charm of the historic courtyard homes teetering precariously on the mountainside, then venture beyond the village where you can access Jingxigudao, an ancient post road, for a nice hike. *¥35; 2hr by bus 892 from Pingguoyuan subway station, then a short taxi ride.*

THE GREAT WALL

The most robust sections of the 21,000km-long Great Wall can be reached in a day or half-day from Běijīng by car. They were built along the ridges of nearby mountain ranges during the Ming dynasty to protect the newly designated capital city. See classic views alongside hordes of visitors at Bādálǐng, take a toboggan ride down at Mùtiányù, explore the wild Wall on a guided hike of an unrestored section, or go by night for one of the music festivals and parties that have sprung up in recent years.

© Sino Images / Getty Images; © Mirko Kuzmanovic / Shutterstock; © Reuben Teo / Shutterstock

07 Shilinxia Glass Platform

This UFO-shaped glass platform opened in 2016 at Jingdong Stone Forest, northeast of Běijīng. The rounded sightseeing platform juts out from the side of a cliff, some 400m above the Shilin gorge. Combine the visit with an hour and a half's vigorous hike up or, for those who favour the more relaxed approach, take a scenic cable car. *www.jdslx.com; 8.30am–5.30pm, closed in winter; 2hr by bus 852 from Dōngzhímén Transport Hub to Pinggu District Bus Terminal, then bus 25 to Pinggu Stone Forest.*

08 Héběi Coast

Find sun, sand, quiet and fresh air in Qínhuángdǎo, in surrounding Héběi province. Nandaihe beach is home to the unexpected Sanlian Public Library: a cool, modern space with huge picture windows overlooking the sea. Běidàihé, meanwhile, was once a popular summer retreat for communist cadre officials and can be combined with a trip to Shānhǎiguān, where the Great Wall meets the sea. *2hr to Beidaihe by train from Beijing Railway Station, then 15min bus to Nandaihe.*

09 Yinshan Pagoda Forest

Hidden away in Běijīng's Chāngpíng district is a set of 18 pagodas that date to the Jin dynasty (1115–1234). Located on the site of the much-earlier Yanshou Temple, the pagodas are said to have housed the remains of the temple's monks. Closed for renovation in 2014, the site reopened in 2018. *8am–5pm; 2–2hr 30min by subway to Xi'erqi station, then a 1hr bus.*

10

© HelloRF Zcool / Shutterstock

07

© Xinhua / Alamy Stock Photo

04

THREE HOURS FROM

10 Shídù

This scenic valley is known for its pointed rock formations, which loom over the Jùmǎ River. There are several satisfying hikes in the area, as well as activities including bamboo-rafting, ziplining and bungee jumping, plus a new glass bridge with dizzying views off the cliffs. 'Shídù' mean '10 crossings' – before modern roads and bridges, the journey along this gorge necessitated that number of river crossings. *2–3hr by train from Běijīng West station.*

11 Wuling Shan

Near to the Simatai and Jīnshānlíng sections of the Great Wall, Wuling Shan offers wonderful mountain views and a variety of hiking trails. You'll want to bed down in one of the 'cocoon' rooms at the luxe Dhawa Jinshanling resort and soothe sore feet with a spa treatment. *www.dhawa.com/jinshanling. html; 2hr 30min–3hr by train or bus to Xìnglóng; an information centre is 50m from the train station. In high season, tourist buses operate, otherwise take a taxi.*

12 Chéngdé

Once a summer retreat for the Qing dynasty emperors, this beautiful city is home to some incredible historical sites and temples built to host dignitaries, including the sixth Panchen Lama from Tibet. The imperial palace of Bìshǔ Shānzhuāng (Fleeing-the-Heat Mountain Villa) features China's largest regal gardens, while the striking red Putuozongcheng Temple is a replica of the Potala Palace in Lhasa. *4hr 30min by train from Běijīng Station.*

THE LOCAL'S VIEW

'Even for a native, it's difficult to keep pace at "Běijīng Speed". Everyone is jetting to a bright future without a break. If I have a minute, I slow down and find inner peace gazing at Fahai Temple's Ming dynasty frescoes, or 1000-year-old Dule Temple in Tiānjīn. If I can't face the traffic on the Fifth Ring Road, I'll stay in the Xisi neighbourhood to read in the courtyard at Zhengyang Bookstore, sip a coffee beside Miàoyīng Temple's white pagoda and stuff my face with Korean-style noodles.'

Mu Yun, editor

ASIA'S BEST BRUS

History is a living thing in Asia, where thousand–year-old temples throng with devotees, and battlefields are shrines for silent contemplation.

Old Goa, India

The grand basilicas that line the sleepy streets of Old Goa are a testament to the ambition of the colonial Portuguese, who envisioned a new Lisbon on the banks of the Mandovi River. *7.30am-6.30pm; 1hr from Panaji by bus.*

Wiang Kum Kam, Thailand

You don't have to go far from Chiang Mai to escape the masses. On the southern limits, the ruins of the original city spill out along quiet country lanes. It's ancient history with a side order of village life. *1hr from Chiang Mai by bicycle.*

Koh Ker, Cambodia

For a taste of the wonder that greeted the first outsiders to lay eyes on Angkor, travel north to Koh Ker, where lavish engravings and Khmer-style towers lie choked in jungle, far from the maddening crowds. *8am-6pm; US$10; 2hr from Siem Reap by car.*

Borobudur, Central Java, Indonesia

Like Angkor Wat in Cambodia, Borobudur is a physical representation of Mt Meru – for Hindus and Buddhists, the spiritual centre of the universe. Its graceful tiers of stupas mirror the volcanoes that rise out of the mist along the horizon. *Magelang; 6am-5pm; US$25; 2hr from Yogyakarta by bus.*

Imperial Tombs, Vietnam

The American Army flattened much of downtown Hue, but the shells didn't get as far as the imperial tombs. Within an hour of the city centre, the elaborate mausoleums of a string of Vietnamese emperors still drip with imperial grandeur. *Duong Xuan Thuong; 7am-5.30pm; 150,000 dong (US$6.50); 1hr from Hue by motorcycle.*

HES WITH HISTORY

Lóngmén Grottoes, China
Despite the ravages of art thieves and Cultural Revolution-era ideologues, the Lóngmén Grottoes outside Luòyáng remain one of the finest collections of Buddhist art in China. Images here range from the miniature to the monumental. *Luolong district, Luòyáng; 8.30am-5.30pm; ¥100 (US$14.50); 3hr from Xi'an by train.*

Ganden Monastery, Tibet
Nowhere captures the losses of the Cultural Revolution and the resilience of the Tibetan community quite like Ganden, whose carefully restored, 15th-century chapels – set in a lonesome location overlooking the Kyichu Valley – still bear the scars of China's 1959 bombardment. *Dagze County, Lhasa; dawn-dusk; ¥50 (US$7.25); 2hr from Lhasa by car.*

Ridi Vihara, Sri Lanka
An easy bus ride from Kandy, peaceful Ridi Vihara was where the silver that paid for the construction of Anuradhapura was mined. Tucked beneath a granite outcrop are cave shrines, shimmering Buddhas, a mini-dagoba (stupa) and South Indian-style temple. *Ridigama; 7am-8pm; 2hr from Kandy by bus.*

Gyeongju, Korea
They don't call Gyeongju the 'museum without walls' for nothing – it must be something to do with the astonishing collection of tombs, palaces and temples left behind by the Silla kingdom, and the lack of walls dividing visitors from all this history. *9am-10pm; from ₩1500 (US$1.35); 2hr from Busan by train and bus.*

Kumano Kodo, Japan
It's not the sights that mark out the Kumano Kodo pilgrimage route, it's the chance to walk in the footsteps of emperors and Samurai, past a string of waterfalls, onsens, ancient shrines and spiritual 'power spots' in the pine-scented hills. *Kii Peninsula, Wakayama; 24hr; 3hr from Osaka by train and bus.*

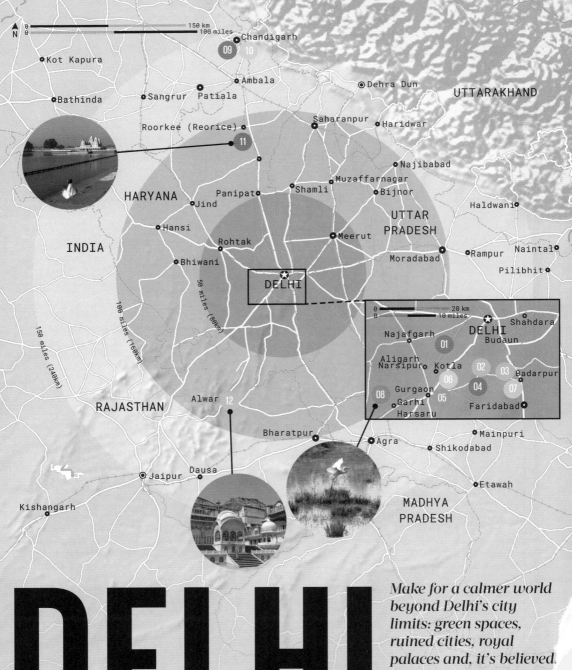

Kot Kapura

Bathinda

Chandigarh

09 10

Ambala

Dehra Dun

UTTARAKHAND

Sangrur Patiala

Roorkee (Reorice)

Saharanpur Haridwar

11

Najibabad

HARYANA

Panipat

Jind

Shamli

Muzaffarnagar

Bijnor

Haldwani

Rohtak

Meerut

UTTAR
PRADESH

Hansi

INDIA

Bhiwani

Moradabad

Rampur

Naintal

Pilibhit

50 miles (80km)

DELHI

100 miles (160km)

150 miles (240km)

0   20 km
0   10 miles

DELHI

Shahdara

Najafgarh

01

Budaun

Aligarh
Narsipur

Kotla

02

03

Badarpur

RAJASTHAN

Alwar

12

06

04

08

Gurgaon

05

07

Faridabad

Garhi
Harsaru

Bharatpur

Agra

Mainpuri

Shikodabad

Jaipur

Dausa

MADHYA
PRADESH

Etawah

Kishangarh

DELHI

*Make for a calmer world
beyond Delhi's city
limits: green spaces,
ruined cities, royal
palaces and, it's believed,
the exact spot where
Brahma created the
universe. Only in India..*

● ARTS & CULTURE　　● HISTORY　　● OUTDOORS　　● FOOD & DRINK　　● FESTIVALS & EVENTS　　● MUSIC & FILM

——— ONE HOUR FROM ———

01 Sulabh Museum of Toilets
Toilets are a regular topic of conversation for travellers to India, and this eccentric museum digs deep into the history of the humble water closet. It's run by an NGO that brings sanitation to India's poorest citizens, and you'll definitely think differently about the country's loos when you realise the hardships the people who clean them have to endure. *www.sulabhtoiletmuseum.org; 10am–8pm, till 5pm Sun; 1hr by metro and autorickshaw from Rajiv Chowk.*

02 Mehrauli Archaeological Park
The lovingly restored Qutb Minar grabs all the headlines, but nearby Mehrauli trades polish for authenticity. Spilling from the forest are monuments spanning 1000 years of history, from the Hindu kingdom of Lal Kot through a succession of Muslim sultanates to the British Raj. Some are still in daily use, including the Hijron ka Khanqah, patron shrine of Delhi's transsexual hijra community. *Dawn–dusk; 1hr by car from Old Delhi.*

03 Tughlaqabad
Eight historic cities grew up on the site of modern Delhi, but Tughlaqabad has the bonus of being a) an atmospheric, overgrown ruin and b) cursed. Founder Ghiyas-ud-din Tughlaq poached construction workers from a Sufi saint, who proclaimed that only shepherds would inhabit the damned city. Today, more goats than humans wander among the stones, while Ghiyas-ud-din slumbers in a forgotten mausoleum. *Dawn–dusk; 1hr by metro and autorickshaw from Jama Masjid.*

04 Chhattarpur Mandir
This suburban temple doesn't draw the crowds like Delhi's landmark Akshardham, but that's part of the appeal. There's a touch of Disneyland about the supersized statues and temple buildings aping every architectural style in southern India, but also a distinctly un-Delhi-like air of calm. Pause under the tower-block-sized statue of Hanuman and slowly exhale. *www.chhattarpurmandir.org; 4am–midnight; 1hr by car from Old Delhi.*

05 Kingdom of Dreams
Beyond the southern suburbs, Delhi-ites are forging a new history for themselves in the satellite city of Gurgaon (Gurugram). Restaurants, bars and clubs buzz nightly, but the big show in town is Kingdom of Dreams, a Vegas-style extravaganza using Bollywood as its source material. After the curtain falls, you can graze around the subcontinent in mocked-up locations from 14 Indian states. *www.kingdomofdreams.in; 1hr by car from Old Delhi.*

06 Amaranta, Gurgaon
A mounted swordfish leaps off the wall at the Oberoi Gurgaon's flagship restaurant – freshly caught seafood, flown in daily from the coast, is the base for many of the kitchen's imaginative modern-Indian creations. Juxtaposition is the name of the game here – think *haleem* (wheat and lamb stew) croquettes and *farsan* (salty snacks from Sindh) served in bento boxes. *www.oberoihotels.com/hotels-in-gurgaon/restaurants/amaranta; 12.30–3pm & 7pm–midnight; 1hr by car from Old Delhi.*

EIGHT KINGDOMS OF DELHI

Delhi is not one city but an amalgamation of eight different capitals, founded by a succession of kings, sultans and emperors. The oldest city of Delhi, Qila Rai Pithora, was a Hindu entity, but Muslim sultans swept into town in the 12th century, founding Mehrauli and the first mosque in India. Rival Muslim empires raised Siri, cursed Tughlaqabad, djinn-haunted Firozabad, and the cultured city of Shergarh, before the Mughals stamped their identity on the city at Shahjahanabad. Last but not least, the British put the icing on the cake with the construction of Edwin Lutyens' immaculately laid-out New Delhi in 1911.

07 Surajkund Crafts Mela

Sun-worshippers have been gathering at this sacred lake to admire the golden reflected rays since the 10th century, but it's in February that Surajkund really comes into its own. The annual crafts fair is a vast gathering of artists, craftspeople, musicians and dancers from across the Indian plains, and a great place to appreciate the remarkable creativity of the subcontinent. *2–18 Feb; 1hr 30min by bus from ISBT.*

08 Sultanpur National Park

Unspoilt nature is probably the last thing you expect to find in Delhi's endless sprawl, but suburban Sultanpur National Park serves up 145 hectares of waterlogged wetlands – a prime stop for migratory birds on the Central Asian Flyway and one of the country's best birding spots. Siberian cranes, storks, ibis and flamingos all put in regular appearances from October to March. *www.haryanaforest.gov.in/en-us/ Wild-Life/Protected-Area/Sultanpur- National-Park-District-Gurgaon; 7am– 4.30pm; 1hr 30min by car from Old Delhi.*

——— TWO HOURS FROM ———

09 Nek Chand Rock Garden

Ride the train out to Chandigarh to marvel at this fantasy garden of concrete and rubbish, built secretly by transport worker Nek Chand. Ceramic herds of cattle cavort with broken bangle dancers, and giant swings dangle beside concrete cascades. Identifying pieces of upcycled junk in the sculptures is just part of the fun. *www.nekchand.com; 9am–6pm; 2hr 30min by train from New Delhi.*

10 Ghazal

Conceived in 1950, with visionary Le Corbusier at the helm, aspects of Chandigarh feel endearingly old fashioned today. At Ghazal, for example, the city's great and good feast on classic Mughlai cooking beneath a vaulted art deco ceiling. It's the kind of place that would have suited Sean Connery's Bond, with a well-groomed waiter guarding a line of single malts at the bar. *SCO 189–191, Sector 17C; 11.30am–11.30pm; 2hr 30min by train from New Delhi.*

© Don Mammoser / Shutterstock

03

——— THREE HOURS FROM ———

11 **Brahmasarovar (Kurukshetra)**
After the chaos and cacophony of Delhi, a short trip out to the country town of Kurukshetra will put you back on the spiritual map. Centred on the Brahmasarovar, India's largest ceremonial tank, this was the spot where Brahma created the universe and where Krishna dictated the Bhagavad Gita. Accordingly, the sacred site is mobbed by pilgrims and holy men. *3hr by car.*

12 **City Palace, Alwar**
A slice of Rajasthani extravagance within reach of Delhi, Alwar's City Palace was once the playground of an eccentric maharaja, but it lives on with faded charm as a dusty complex of government offices. At the heart of the palace is a ceremonial tank reflecting a quadrangle studded with pavilions, turrets and ghats, where the eccentric owner used to come to contemplate the spheres.
3hr by train from New Delhi.

<u>THE LOCAL'S VIEW</u>

'My all-time favourite escape from Delhi, especially in winter, is the Chambal Safari Lodge near Bah. Its tranquil and verdant ravines are filled with birdsong and resident wildlife, providing succour to my din-doused soul. The lodge lays on a sumptuous lunch spread of regional delicacies, and it's set within a three-hour drive from Delhi, via two of India's best expressways, making it even more inviting as a day trip. When time allows, I drop in at the nearby temple town of Bateshwar on my way back to the capital.'

Puneetinder Gaur Sidhu, writer

07

© India Picture/UIG / Getty Images

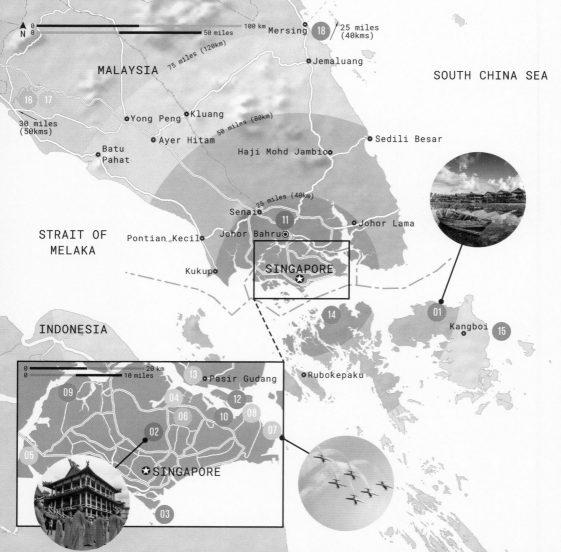

Central Singapore is packed with skyscrapers and heaving lanes but on the island's outskirts it's a different city. Local temples, natural wonders and gastronomic delights abound, and in nearby Malaysia and Indonesia there's more to discover.

SINGAPORE

● ARTS & CULTURE ● HISTORY ● OUTDOORS ● FOOD & DRINK ● FESTIVALS & EVENTS ● MUSIC & FILM

——— ONE HOUR FROM ———

01 Pulau Bintan

This charming Indonesian island is a favourite weekend spot. The beaches closest to Singapore are dotted with fancy resorts, but further afield you'll find local towns with plenty to see. Traverse a floating Chinese village, visit a Buddhist temple swallowed by a Bayan tree or take to the sky kitesurfing. Ferries depart Tanah Merah Ferry terminal, and once on the island taxis and private cars can be hired. *1hr by ferry.*

——— TWO HOURS FROM ———

02 Kong Meng San Phor Kark See Monastery

Recentre your mind and soul at Singapore's largest Buddhist monastery. Time easily slips away as you wander the ornately decorated temples, tranquil manicured gardens peppered with koi ponds, and meandering pathways. Be awed by the 13.8m bronze Buddha in the Hall of No Form and take time to reflect under the bodhi tree. Dress appropriately with knees and shoulders covered. *1hr 10min by bus 410 or 52 from Bishan MRT station.*

03 The Southern Islands

Escape the shiny city to this near-deserted trio of offshore islands. Alight at St John's and cross the concrete walkway to Lazarus, to frolic on arguably Singapore's most picturesque beach. Sunbathing done, head to Kusu and climb the 152 steps to the 19th-century Malay shrines. Be sure to bring plenty of water and food. *Ferries (www.islandcruise.com.sg) depart Marina South Pier, accessible via the Marina South Pier MRT station. 1hr 10min to St John's from the CBD.*

04 The Summerhouse Garden Dome Dining

Singapore's hot and humid weather can make alfresco dining a slightly sweaty affair, but that's all changed with The Summerhouse's three cooled dining domes. Located in the restaurant's edible garden, guests can pick from three interior themes, Bohemian, Scandinavian and Lounge, and relax with mood lighting, piped music and delectable farm-to-table cuisine. *www.thesummerhouse.sg/garden-domes; 1hr 20min by MRT to Khatib station, then bus 117.*

05 Tiger Beer Brewery Tour

Sadly, you won't find any wild tigers stalking the rainforests in Singapore; the last one was shot in 1930. It's little compensation, but you will see a tiger emblazoned on every bottle of Tiger Beer, brewed in the Lion City since 1932. You can delve into the brand's history and brewing methods at the Tiger Beer Brewery. Guests over 18 get to sample the golden ale! *www.tigerbrewerytour.com.sg; 1hr 30min by MRT to Joo Koon station, then bus 182.*

06 Lorong Buangkok

Step back in time at Singapore's last remaining *kampong* (village). The pace is slow here, with crickets humming, chickens roaming about and ramshackle timber houses peeking from behind overgrown pathways. *1hr 30min by bus 88 from Ang Mo Kio MRT station to Ang Mo Kio*

HERE, THERE, EVERYWHERE

Singapore's sleek and efficient MRT system is not only the easiest, quickest and coolest (all carriages are air-conditioned) way to get around this city-nation, but it's also the cheapest. Single trip fares cost between just S$1.40–2.50 and even less when using an EZ-Link card, available for purchase at MRT stations and 7-Elevens. Once your card is topped up with credit, you can quickly 'tap' on and off as you commute; the cards are also accepted on city buses.

© Jonathan Siegel / Getty Images

Ave 5, then walk north up Yio Chu Kang Rd and turn right into Gerald Dr. Lorong Buangkok is 200m further, follow the dirt track on your left.

07 Singapore Airshow

Featuring death-defying aerobatic displays and the chance to get up close and personal with pilots and aircrafts, this aerospace festival is an aviation fanatic's heaven. The show is held once every two years and tickets are snapped up quickly, so keep your ear to the ground for release dates. *www.singaporeairshow.com; Feb biennially; 1hr 30min by MRT & shuttle bus.*

08 Changi Village Hawker Centre

Foodies should head straight to this seaside hawker centre, which serves up some of Singapore's best local grub. You'll find everything from chicken rice to stir-fried noodles, plus two famous nasi lemak (coconut rice topped with fried chicken or fish, fried anchovies and sambal chilli) stalls (01-03 and 01-26) continually duelling to produce the best dish. *1hr 30min by MRT to Simei station, then bus 9.*

09 Kranji Countryside

Land in Singapore is in seriously short supply, so it's surprising to find some farms still thriving in the northwest. Bollywood Veggies bursts with tropical plants and fruit trees, and makes an excellent lunch spot, while kids can feed goats at Hay Dairies and get up close to American bullfrogs at Jurong Frog Farm. Hop on the handy daily minibus service (www.kranjicountryside.com) running from Kranji MRT station. *1hr 30min by MRT & Kranji Countryside Express.*

10 Kayak Fishing

It's just you, rod, pedal-powered kayak and guide (www.fever.sg) as you take to the waters in search of the perfect fishing spot. You'll glide through mangrove swamps, where dinosaur-like monitor lizards lurk in the undergrowth, and quaint *kelongs* (traditional floating homes and fisheries) while learning the area's history and trying to hook a catch. *1hr 30min by MRT to Pasir Ris station & a short walk to Pasir Ris Park.*

11 Johor Bahru

Just over the causeway in Malaysia is bustling border town Johor Bahru. JB's chaotic and gritty atmosphere is in sharp contrast to gleaming and orderly Singapore, but that's its appeal. Spend the day wandering the heritage district lanes, many decorated with colourful street art, munching on local foods and visiting the area's temples and museums. *1hr 30min by express bus from Queen Street Terminal, but allow longer on weekends.*

12 Pulau Ubin

Outdoorsy adventurers will love this island off northeast Singapore. Mountain bikers peddle straight for Ketam Mountain Bike Park, while nature enthusiasts can ramble through Chek Jawa Wetlands. Singaporeans get nostalgic for yesteryear in the old *kampong* (village) while tucking into chilli crab by the seaside. Overnight camping is permitted. *2hr by MRT to Simei MRT station, then bus 9 to Changi Village & 10min bumboat trip.*

13 Pasir Gudang World Kite Festival

There's something magically nostalgic about a blue sky flecked with colourful, fluttering kites. It's not all fun and games at this annual festival in Pasir Gudang,

though, as flyers from all over the world converge to unveil their creations and battle kite to kite in fierce contests. Bystanders can take to the air with their own kites and visit the informative kite museum. *www.kitefestpasirgudang.com; Feb; 2hr by car.*

14 Pulau Batam

Luxury spas and rounds of golf don't come cheap in the Lion City, unlike much of Southeast Asia. Instead of maxing out your credit card, jump on the direct one-hour ferry from HarbourFront to neighbouring Batam, in Indonesia, to enjoy indulgent pampering and golf-ball whacking. Stay overnight at Tempat Senang (www.tempatsenang.com), where the in-house spa is one of the best. *2hr 30min by MRT & ferry.*

THE LOCAL'S VIEW

'Mountain-bike riding is my way of escaping the urban jungle. There are quite a few trails around Singapore; my favourite is at Chestnut Park. However, if I really want to get away, I take my bike on the bumboat over to Pulau Ubin and spend a few hours going hell for leather on the gruelling trails at Ketam Mountain Bike Park. I finish my ride with an ice-cold coconut juice – straight from the nut.'

Jason Tan, web designer

© Courtesy of Fever

——— THREE HOURS FROM ———

15 Nikoi Island Resort
It's the Robinson Crusoe Island of your dreams. Swaying palms, driftwood huts, golden sand and a sparkling sea, complemented by five-star service, Indonesian dining and beach cocktails. Kids can go wild on resort-led treasure hunts, jetty jumping competitions and tennis tournaments. The young at heart are welcome to join in too. *www.nikoi.com; 2hr 30min by ferry from Tanah Merah terminal, then shuttle & boat as arranged by resort.*

16 Melaka
Heaving with heritage sites, a result of both its rich Portuguese, Dutch and British colonial history, and Malay, Chinese and Indian cultural mix, Melaka is a place you'll love getting lost in. The Baba and Nyonya Heritage Museum is not to be missed, nor are the gastronomic delights; a fusion of all the nationalities that have called Melaka home. *3hr by bus.*

Melaka Art & Performance Festival

Colonial town Melaka comes alive with contemporary culture and performances during this annual festival held each November. Expect the unexpected as artists from around the world descend on this Unesco World Heritage–city, ready to express themselves through dance, music, workshops, exhibitions and theatre. Don't think you'll be sitting in an air-conditioned theatre, however, as many of the performances are held throughout the city's historical site.
www.melakafestival.com; Nov; 3hr by bus.

18 **Pulau Tioman**

Lush, wild and teeming with wildlife, Tioman Island acts as a beacon for nature lovers and offers every possible shade of paradise. Trek through the jungles to find hidden waterfalls or relax on deserted white beaches fringed by laid-back villages. If you can't sit still, take a boat to go island-hopping, and snorkel or dive among the surrounding reefs and wrecks. As the sun sets enjoy a cold drink and relax till the stars come out. *3hr by car to Mersing jetty then 1hr 30min by ferry.*

SINGAPORE'S SLEEPING DRAGON

In Singapore's western jungles, among a labyrinth of shelved pottery, lies a sleeping dragon. The oldest and only operational kiln of its kind in Singapore, this three-quarter-century-old dragon roars into action three to four times a year (www.thowkwang.com.sg). At 27m in length, the kiln can hold thousands of pieces of pottery and is built on a gentle slope with its fire-breathing mouth at the hill's base and the chimney at the tail. When the dragon is quiet tours are conducted of its inner belly, and you can even make a piece of pottery to be fired during its next awakening. It's a 90-min trip by MRT to Boon Lay station, then bus 199.

© MissHabanita / Getty Images

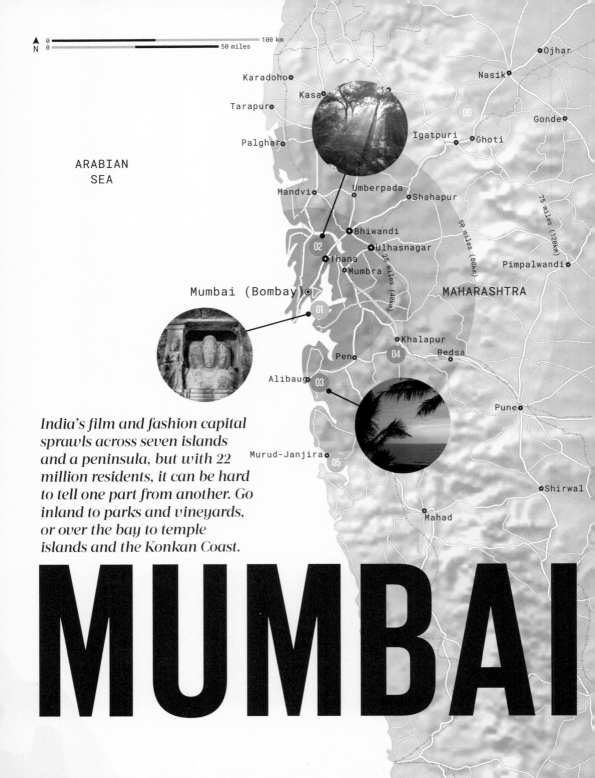

ARABIAN
SEA

MAHARASHTRA

Mumbai (Bombay)

India's film and fashion capital sprawls across seven islands and a peninsula, but with 22 million residents, it can be hard to tell one part from another. Go inland to parks and vineyards, or over the bay to temple islands and the Konkan Coast.

MUMBAI

⬤ ARTS & CULTURE   ⬤ HISTORY   ⬤ OUTDOORS   ⬤ FOOD & DRINK   ⬤ FESTIVALS & EVENTS   ⬤ MUSIC & FILM

─── ONE HOUR FROM ───

01 **Elephanta Island, Mumbai Harbour**
Sure, it's up there on the list of must-see Mumbai sights, but the Gateway of India also offers an escape route, by boat, from the heat and noise of the city. When you've had your fill of the carving-filled cave temples that made Elephanta famous, wander across the island in peaceful solitude to old brick stupas and quiet fishing villages. *9am–5pm Tue–Sun; 1hr by boat from the Gateway of India.*

02 **Sanjay Gandhi National Park, Borivali**
Sanjay Gandhi is a proper national park, right down to the chattering monkeys and stalking leopards who mosey around within striking distance of Bollywood's biggest film studios. Tucked away in the middle of it all is one of Mumbai's most sublime locations: the Kanheri caves, a complex of monasteries, dwellings and halls hollowed out by Buddhist monks in the first century BC. *7.30am–6pm Tue–Sun; 1hr by train & rickshaw from Churchgate station.*

─── TWO HOURS FROM ───

03 **Alibaug Beach**
The sun sets spectacularly over the sand at Girgaum Chowpatty and Juhu Beach, but the waters are polluted this close to downtown. Savvy sea-lovers take a ferry from the Gateway of India to Mandwa and then the bus to Alibaug, where ruined forts tumble down to clean, palm-backed beaches. *2hr by boat from Gateway of India, then bus.*

─── THREE HOURS FROM ───

04 **Adlabs Imagica, Khopoli**
Adlabs Imagica is the premier playground in town – a theme park on an American scale but with Indian overtones, so the 40m rollercoaster is joined by a Bollywood hall of fame and an indoor snow-park, where Mumbaikars come to experience snowfall for the very first time. *www.adlabsimagica.com; 10.30am–8pm; 2hr 40min by train from Chhatrapati Shivaji Maharaj Terminus.*

05 **Janjira Fort, Murud-Janjira**
The Konkan Coast is studded with forts, which were built by everyone from Maharashtran sultans to the British and Portuguese, but topping the romance stakes is this floating sea-fortress, half a kilometre offshore from the fishing village of Murud. Jaunty sailboats cruise out daily, leaving visitors to wander around battlements that rise sheer out of the Arabian Sea, awash with the ghosts of vanquished empires. *7am-dusk; 3hr by car.*

06 **Grover Zampa Vineyard**
Indian wine started out small, but now the whole subcontinent is catching on, thanks to the vineyards covering the hills around Nasik. Pioneering grower Grover Zampa harnesses the local terroir (ideal for cabernet, shiraz and sauvignon blanc) and tasting tours show how far Indian wine has come since the owners first started cultivating imported French vines for juice in 1992. *www.groverzampa.in; tours at 10.30am, 2.30pm, 4pm; 3hr by car.*

THE LOCAL'S VIEW

'After a long work day, I like to unwind with a cup of herbal tea in one of Mumbai's amazing tea cafes. But when I'm in a more indulgent and adventurous mood, I join a guided walk through one of the forest areas around the city, arranged through the Bombay Natural History Society, one of India's longest established nature conservation organisations. To get a good tan, I head to the beaches on the outskirts of Mumbai; one of the best is just an hour away from Bandra on Manori island.'

Deepika Gumaste, blogger & writer

▲ N 0 ———— 50 km
     0 ———— 32 miles

NORTH KOREA

KOREAN DMZ (NORTH)
KOREAN DMZ (SOUTH)

⊙ Kaesong

WEST SEA (YELLOW SEA)

48 miles (75km)

32 miles (50km)

Chuncheon ⊙

16 miles (25km)

09
10
06
Goyang ⊙
04 02
07
★ SEOUL
12

11
Incheon ⊙
Puch'on ⊙
Seongnam ⊙
08
Wonju ⊙

11

Ansan ⊙
01
Suwon ⊙
03
05
Eumseong

SEOUL

Seoul is a 24-hour city with great food and nightlife. But if you tire of hyper-technology and neon lights, retreat to rugged parks, quiet villages, or the fearsome Demilitarized Zone.

● ARTS & CULTURE   ● HISTORY   ● OUTDOORS   ● FOOD & DRINK   ● FESTIVALS & EVENTS   ● MUSIC & FILM

———— ONE HOUR FROM ————

01 Suwon

Suwon almost became the country's capital in the 18th century, when Joseon dynasty ruler King Jeongjo built 5.7km-long city walls ahead of moving the royal court south. The king's death meant power stayed in Seoul, but the Unesco-listed fortress remains. A circuit of the fortifications takes you past ancient gates, command posts, pavilions, observation towers and fire-beacon platforms. ***30min–1hr by high-speed train or Budang Line subway.***

02 Bukhansan National Park

Hike through granite-studded peaks and explore hidden temples in this 80 sq km national park, easily visitable via Seoul's sprawling subway. A popular option is to clamber up the ridgetop path to Dobong-san (740m), but if you're up for a challenge tackle South Korea's highest peak, Baegundae (836m), or go rock climbing on Insu-bong (810m). Or simply sit back with a picnic by a mountain stream. ***http://english.knps.or.kr; 45min by subway to either Gupabal or Gireum stations.***

03 Everland Resort

Hit water-park waves, idle along a lazy river or scream your way down South Korea's best rollercoaster (the T Express, with a 45m, near-vertical drop) at the country's biggest theme park. There are five zones of rides and a series of serene gardens. In the evenings, a fireworks display is set dramatically against the surrounding rolling hills. ***www.everland.com; 9.30am–10pm; 50min by express bus from Everland Bus Stop in Gangnam.***

04 Eunpyeong Hanok Village

Built in 2016 as an alternative to similar over-touristed *hanok* (traditional tiled-roofed house) villages, this neighbourhood was designed using traditional methods and offers respite from the modern city and a glimpse of a simpler way of life. Some *hanok* here offer accommodation, others teahouses or cafes, and some are private residences. ***http://museum.ep.go.kr; 9am–6pm; 30min–1hr by subway to Gupabal station.***

05 Icheon Ceramic Village

The village of Icheon has been a pottery centre since the Joseon dynasty (1392–1897), and you can check out historic and modern versions of the craft in this enclave of workshops and retailers. There are ceramics shops, kilns and cafes, and workshops where you can fire your own creations. ***www.icheon.go.kr; 10am–5pm; 1hr by bus from Dong-Seoul Bus Terminal.***

06 Forest of Wisdom

This library contains some 200,000 books and reading materials and a calming cafe. It's located in Heyri Art Valley, a laid-back town filled with similar spaces: galleries, award-winning buildings, street art and cafes, all conceived around Paju Book City, a neighbourhood dedicated entirely to about 250 publishing companies. ***www.forestofwisdom.or.kr; 10am–5pm; 1hr 20min by express bus from Hapjeong subway station.***

07 Geumsun-sa Templestay

Spend a blissful weekend or even a whole week away at this 600-year-old

THE DMZ

Though it may not be the most typically relaxing day trip, a visit to the highly fortified border with North Korea offers a chance to set foot inside the Joint Security Area, a military zone where the Korean Armistice Agreement was signed in 1953. Today, visitors tour Panmunjom 'Truce Village', crossing briefly into North Korea. The DMZ is a one-hour drive from Seoul and trips must be arranged through a tour agency; most are a half-day long and include visits to Dora Observatory (where you can gaze into North Korea through binoculars) and to some of the infiltration tunnels dug under the border.

Buddhist temple in Moaksan Provincial Park in the mountains north of Seoul. The guest programme includes learning meditation, prostrations, craft-making, tea ceremonies and forest walks in the company of resident monastics. ***http://eng.templestay.com; 1hr 30min by subway.***

 08 **Museum SAN**
Set in lush mountains near the town of Wonju, this museum is sensitively designed to harmonise with its surroundings and function as a place of peace and calm. There are flower, water and stone gardens outside, and inside exhibits explore natural elements, such as paper, light and space. ***www.museumsan.org; 10.30am–6pm, closed Mon; 1hr 30min by train from Cheongnyangni station, then shuttle bus.***

——— **TWO HOURS FROM** ———

 **09** **Goryeogungji Palace**
The remains of this small palace sit on a hillside on the island of Ganghwado – construction was completed around 1234, soon after King Gojong moved his capital to Ganghwado to better resist Mongol invasion. In 1866, it was destroyed by invading French troops and has since been partially rebuilt and three major gates have been renovated. It offers a fascinating insight into this niche of history. ***9am–6pm; 1hr 50min by bus from Sinchon station.***

10 **Joyang Bangjik**
Head out to one of South Korea's largest islands, Ganghwado, to this textile factory turned cafe. You can grab a coffee, beer or a fresh juice and plonk down on one of the long, rustic wooden benches inside the giant skylit main hall. Later, explore the maze of different rooms and spaces, all kitted out with shabby-chic tables and chairs, across this rambling complex. ***1hr 50min by bus from Sinchon station.***

11 **Yeongjongdo & Muuido**
This pair of islands provide wonderful beachside escapes (as well as international escapes via Incheon Airport). Yeongjongdo's

06

WINTER SPORTS

Winters in South Korea can be frigid, with temperatures below zero. The peninsula's chill creates some great getaways: ski resorts such as Elysian Gangchon, Bears Town, Yongpyong and Alpensia (venue for the 2018 Winter Olympics) are reachable from Seoul in a couple of hours by bus or train. Likewise, Koreans embrace the cold by staging festivals of ice-fishing, skating and sculpture-making to a K-Pop soundtrack. In January, head to Hwacheon Sancheoneo Ice Festival (2hr 30min by train from Cheongnyangni) or further afield to Taebaeksan Snow Festival (3hr by bus from Dong-Seoul or 4hr by train from Cheongnyangni).

Eulwangni is the most accessible beach from Seoul and features cubist-meets-modernist retreat Nest Hotel, while less-developed Muuido is your best bet for beach huts, camping and seafood on the sand. *1hr 30min–2hr by train and ferry from Seoul Station via Incheon Airport.*

12 Myeongji Valley

This river valley in Gapyeong, northeast of Seoul, is full of cascading waterfalls and stunning scenery. Pitch up at a glamping site, go apple-picking or sample *galbi* (beef short ribs) in Pocheon Idong, a whole village dedicated to the dish. *1–2hr by subway to Gapyeong, then local bus.*

12

Koriyama
120 miles (195km)

18

09

Iwaki

12

80 miles (130km)

06

Nagano

Utsunomiya

15

Mito

Maebashi

Takasaki

40 miles (65km)

Matsumoto

07

02

01

Saitama

Kawaguchi

08

11

★ TOKYO

03

Chiba

Kawasaki

14

Yokohama

13

05

04

Yokosuka

10

Fuji

17

Shizuoka

16

Hamamatsu

PACIFIC OCEAN

TOKYO

Even Tokyo eventually gives way to forests, mountains and sea. Outside the city, the link to the past is palpable, yet the countryside has also inspired many modern artistic projects.

ARTS & CULTURE   HISTORY   OUTDOORS   FOOD & DRINK   FESTIVALS & EVENTS   MUSIC & FILM

——— ONE HOUR FROM ———

01 Kawagoe's Ko-edo

'Ko-edo' means 'little Edo' – Edo being the former name for Tokyo, in use until the mid-19th century. Tokyo itself has very few buildings that recall the days of Edo, but Kawagoe's ko-edo district has managed to hold on to several, including whitewashed, mud-walled warehouses, low-slung shops with tiled roofs and sloping eaves, and the town's signature wooden watchtower. *www.koedo.or.jp; 30min by train from Ikebukuro.*

02 Omiya Bonsai Art Museum

Thanks to a lucky confluence of historical and ecological factors, the otherwise ordinary northern suburb of Omiya evolved into Japan's bonsai capital. The excellent Omiya Bonsai Art Museum exhibits over a hundred examples by masters of the form, with displays that demonstrate different styles and techniques. Within walking distance of the museum are several public bonsai gardens as well as nurseries. *www.bonsai-art-museum.jp/en; 9am–4pm, closed Thu; 40min by train from Shinjuku.*

03 Ishikawa Brewery

It's the water that gives sake its unique characteristics and Ishikawa draws its essence from the Tama River, next to the brewery. The sake here is made with traditional methods and local ingredients – not so different from when the brewery started out in 1863. Book ahead for a free, 90-minute English-language tour of the 19th-century facilities; sake tasting included. *www.tamajiman.co.jp/en; 10am–4pm Mon-Fri; 1hr by train from Shinjuku.*

04 Kamakura

Kamakura had a brief turn as the capital of Japan back in the 12th and 13th centuries – which is also when Zen Buddhism entered Japan. As a result, Kamakura has dozens of Zen temples, distinguished by their severe beauty, heavy upturned roofs and monumental wooden gates. Some temples, including Kenchō-ji and Enraku-ju, host public *zazen* (seated meditation) sessions – the ultimate antidote to hectic city life. **1hr by train from Shimbashi.**

05 Tanabata Festival, Hiratsuka

Tanabata is a celebration of lovers and light, held annually on 7 July – the day that the stars Altair and Vega (stand-ins for two star-crossed lovers) meet across the Milky Way. Hiratsuka, on the Shōnan Coast, hosts a big street party for the occasion (over the weekend that falls nearest to 7 July) with gorgeous handmade lanterns, plenty of food stalls, and couples strolling side-by-side in colourful summer kimonos. *www.tanabata-hiratsuka.com; 7 Jul; 1hr by train from Tokyo Station.*

06 Water Garden, Art Biotop

To create this enchanting garden, experimental architect Ishigami Junya repositioned hundreds of native trees around miniature reflecting pools, with unobtrusive walking paths and velvety moss in between. It forms part of the Art Biotop complex, which also has a farm-to-table restaurant and a gallery, in the popular resort area of Nasu-Shiobara. *www.artbiotop.jp; tours daily; 1hr 10min by bullet train from Tokyo Station, plus free shuttle bus.*

ALL ABOARD

The JR Tokyo Wide Pass (adult/child ¥10,000/5000; www.jreast.co.jp/e/tokyowidepass) covers train travel for three consecutive days between central Tokyo and some of the capital's more far-flung getaways. This includes rides on Tobu's limited express train to Nikkō and Izukyū's limited express train to Shimoda. It also covers the shinkansen (bullet train) as far as Echigo-Yuzawa (where you can rent a car for a day to explore the Echigo-Tsumari Art Field), Nasu-Shiobara and, during ski season, Gala Yuzawa. Buy at any JR East Travel Service Centre in Tokyo or at Narita or Haneda airports.

07 Chichibu Night Festival

Chichibu's annual night festival is one of Japan's biggest and most famous *matsuri* – the traditional festivals that have roots in the country's indigenous Shintō religion. Enormous, shrine-like floats, lantern-lit and with cusped roofs, are pulled through the streets by hand. Further spectacles include traditional dancing and music, and a dramatic finale of fireworks. *www.chichibu-jinja.or.jp; 2 & 3 Dec; 1hr 20min by train from Ikebukuro.*

08 Narita Drum Festival

Narita's signature temple, Shinshō-ji – centuries-old and always smoky with incense – is impressive in its own right; when traditional drum teams from all over the country show up for one weekend in April to play on the grounds, it's spectacular. There are athletic performances on the *taiko*, Japan's deeply resonant 'big drum'; a thousand-drummer strong 'Prayer for Peace'; and a torchlit night show featuring the top teams. *www.nrtm.jp; 1hr 20min by train from Ueno.*

09 Snowsports, Gala Yuzawa

Tokyo's closest winter resort is custom-made for a populace enamoured with convenience: it has its own shinkansen (bullet train) station connected to the gondola station (where you can rent everything you could possibly need). All that and decent powder, with runs suitable for all levels. The season at Gala varies but is generally long, from mid-December to April. *https://gala.co.jp/winter/english/index. html; 8am–5pm; 1hr 20min by bullet train from Tokyo Station.*

© Marek Slusarczyk / Alamy Stock Photo

© Magda Rittenhouse / Alamy Stock Photo

10 Enoura Observatory

Perched as it is on the edge of the Hakone Mountains, with unfettered views cascading towards Sagami Bay, an 'observatory' is one way to describe this work by Japanese contemporary artist Sugimoto Hiroshi. But that doesn't take into account the gallery, performance space (check the website's schedule), landscape garden and restored traditional teahouse – all overseen by the genre-defying artist. *www.odawara-af.com/en; admission by advanced reservation only; 1hr 30min by car.*

11 Oku-Tama

The Oku-Tama is Tokyo's natural playground (thanks to an administrative quirk, this lushly forested region of peaks and rivers is technically part of Tokyo). The classic hike here starts in the mountains at Musashi Mitake-jinja (accessed via cable car) and wends further upwards through cedar groves to Ōtake-san – where you might get a view of Mt Fuji – and back (a five-hour loop). *www.okutama.gr.jp; 1hr 30min from Shinjuku to Mitake.*

⎯⎯ TWO HOURS FROM ⎯⎯

12 Tōshō-gū, Nikkō

Nikkō is a national park in the mountains north of Tokyo; Shintō shrine Tōshō-gū holds the deified remains of one of Japan's great historical figures, the shogun Tokugawa Ieyasu. Shrines in Japan are often humble structures, but not this one: the most skilled artisans of the 17th century were called upon to create lavishly ornate structures. The backdrop of towering cedar trees adds atmosphere in spades. *www.toshogu.jp; 8.30am–5pm; 1hr 50min from Asakusa.*

11

13 Amazake-chaya

Amazake is a naturally sweet, non-alcoholic drink that's made from fermented rice. At thatched-roof teahouse Amazake-chaya the experts have been making it exactly the same way for nearly four centuries. The shop is situated along the Old Tokaidō Highway, which connected Tokyo and Kyoto during the Edo period. The poor porters who used to run this route during that time loved the stuff for its fortifying qualities. *www.amasake-chaya. jp; sunrise to sunset; 2hr 30min by train & bus from Shinjuku.*

14 Climbing Mt Fuji

Summiting the country's tallest peak, 3776m-tall Fuji-san, is the ultimate Japan bucket-list activity. It's a crowded, minimum five-hour trek to the top, but one that traces the footsteps of pilgrims from centuries past who worshipped this (currently dormant) volcano. Watching sunrise from the top is a transcendental experience. Climbing season runs from 1 July to 10 September. *www.fujisan-climb.jp; 2hr 30min by direct bus to the trailhead from Shinjuku.*

© iamlukyeee / Shutterstock

14

—— **THREE HOURS FROM** ——

15 Mashiko Museum of Ceramic Art

Pottery centre Mashiko was integral to Japan's early 20th-century arts and crafts movement. Outstanding examples of the earthy pieces produced here (which influenced western potters, such as Bernard Leach) are on display at the museum, as well as a 19th-century traditional climbing kiln. But Mashiko isn't a historical site: around town are dozens of active studios, galleries and pottery shops. Rent a bicycle to see them all. *www.mashiko-museum.jp/en; 9am-4pm; 2hr 40min by bus from Akihabara.*

16 Shimoda

At the tip of the Izu Peninsula, port town Shimoda played a crucial role in Japanese history: US gunboats famously arrived here in the 1850s, demanding that Japan open up to foreign trade (and setting off a cascade of political and cultural shifts). There are plenty of small museums and historical sites to check out, but also a breezy seaside vibe and delicious seafood to enjoy. *2hr 50min by train from Tokyo Station.*

Shira-hama Beach

The name's a giveaway: Shira-hama means 'white-sand beach' and this is the prettiest little stretch of coast within easy access of the capital. On a rocky outcrop sits a bright red *torii* (shrine gate) that makes for a perfect photo-op against the deep-blue waters. The beach is only lightly developed; go outside of school holiday season and you might have it to yourself.

3hr by train & bus from Shinagawa.

 Echigo-Tsumari Art Field

Japan's central mountain spine is the country's great divider; the Echigo region, on the other side to Tokyo, is deep country – all rice paddies and wooden farmhouses with dramatically sloping roofs (to handle all that snow). Time would have buried it all were it not for the creation of this sprawling, ambitious art project that includes site-specific installations from the likes of James Turrell.

www.echigo-tsumari.jp/eng; 3hr by car.

THE LOCAL'S VIEW

'I like to go trekking, to get out into nature. There are many mountains around Tokyo that are accessible by public transportation. Mt Takao is famous, but it gets too crowded. There are other, lesser-known places I like better. Some of the spots I recommend include Mt Kōbō, which is an easy three-hour hike, and Mt Ōyama, which is a more challenging five- or six-hour hike. In both cases you can end the hike with a soak in a local onsen.'

Mako-san, who runs Levain, a Shibuya bakery

Tunnel of Light / MAD Architects / Ma Yansong / Yoshifumi Moriya / Nacasa&Partners

JAPAN'S BEST O

NSEN RETREATS

A trip to Japan wouldn't be complete without a naked soak in a natural hot spring. Note, some onsens don't allow tattoos, and while most baths are gender-segregated, some of the best aren't.

Noboribetsu Onsen

This is among the most onsen-y of all onsen: the whole town smells of sulphur (that's a good thing: Noboribetsu is routinely voted one of Japan's best hot springs). Locals call the raw, volcanic landscape here 'hell valley'. *Noboribetsu; www.noboribetsu-spa.jp; segregated; 1hr 30min from Sapporo by bus.*

Sukayu

Isolated in the mountains of Japan's deep north, Sukayu's 300-year-old cedar wood '1000-person bath' – it's huge! – is a pilgrimage spot for bathing connoisseurs from all over Japan. *Sukayu-zawa; www.sukayu.jp; mixed; 1hr 20min from Aomori by bus.*

Tsuru-no-yu

With open-air baths, milky-white waters and nothing around but the onsen's 17th-century ryokan (traditional inn) and the forested hills beyond, Tsuru-no-yu is pretty much the platonic ideal of the remote rural hot spring. *Semboku; www.tsurunoyu.com; mixed; 1hr 40min from Akita by car.*

Hakone Yuryō

Hakone is Tokyo's favourite retreat, an upscale hill station with seven distinct hot springs. Day spa Hakone Yuryō has outdoor baths with mountain views, and even some that can be booked privately. *Hakone; www.hakoneyuryo.jp; segregated; 1hr 30min from Tokyo by train.*

Hōshi Onsen Chōjūkan

The beautiful, late 19th-century wooden bathhouse here, weathered just right, with exposed beams and latticework on the windows, will make you swoon even before you've hit the waters. *Minakami; www.hoshi-onsen.com; mixed; 2hr 30min from Tokyo by car.*

Awa-no-yu

Rich mineral content makes the waters here opaque – good news for timid bathers. The rocky, leaf-fringed outdoor pool at this ryokan looks like something the gods created for their own pleasure. *Azumi; www.awanoyu-ryokan.com; mixed; 1hr from Matsumoto by car.*

Kinosaki Onsen

Kinosaki is the Kansai region's signature onsen resort town. Don a yukata (a cotton, kimono-like robe) as you hop from bath to bath – there are seven public baths here. *Toyooka; www.visitkinosaki.com; segregated; 2hr 30min from Kyoto, Osaka or Kobe by train.*

Lamp-no-yado

In a private cove at the tippy top of the remote Noto Peninsula, this trad-meets-modern luxury ryokan has dramatic ocean views from the baths and decadent seafood dinners, with ingredients sourced from nearby waters. *Noto Peninsula; www.lampnoyado.co.jp; segregated; 2hr 30min from Kanazawa by car.*

Kurokawa Onsen

Highly volcanic Kyūshū has several famous onsen resort towns, but Kurokawa – holed up in the mountains – is the best, thanks to purposefully limited development. Ideally, you should splurge on one of the ryokan here (all in tasteful, traditional style). *Kurokawa; www.kurokawaonsen.or.jp; segregated; 2hr 30min from Kumamoto by car.*

Ibusuki Sunamushi

Swap water for earth at this coastal spot where strong-armed locals bury you up to your chin in naturally hot sand. You get to wear rented cotton pyjamas, and, oh, will you sweat. *Ibusuki; www.sa-raku.sakura.ne.jp; mixed; 1hr from Kagoshima by train.*

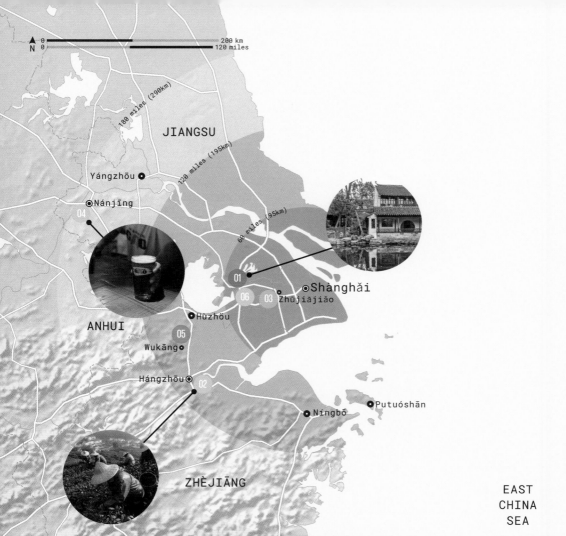

JIANGSU

180 miles (290km)

Yángzhōu

Nánjīng

120 miles (195km)

60 miles (95km)

04

01

06 03 Zhūjiājiǎo

Shànghǎi

Hùzhōu

05

Wukāng

Hángzhōu 02

ANHUI

Níngbō Putuóshān

ZHÈJIĀNG

EAST
CHINA
SEA

Temper Shànghǎi's excess and hectic pace with leisurely getaways: stroll along canals in the water towns of Jiāngsū and Zhèjiāng; enjoy Chinese arts in Sūzhōu; get spiritual at a Buddhist retreat; or explore paths through tall bamboo groves.

SHÀNGHǍI

● ARTS & CULTURE   ● HISTORY   ● OUTDOORS   ● FOOD & DRINK   ● FESTIVALS & EVENTS   ● MUSIC & FILM

——— ONE HOUR FROM ———

01 Garden of the Master of the Nets

Neighbouring city Sūzhōu has more than a dozen classical gardens to visit, but choose the petite Garden of the Master of the Nets for traditional Chinese music performances on warmer evenings. It's hard to resist the magical atmosphere as *èrhú*, *pípá* and *gǔzhēng* (stringed instruments) are played and opera performed, with the moonlight glinting on the garden's shimmering ponds. *7.30am–5pm, evening performances 7.30–9.30pm Mar–Nov; 30min by train from Shànghǎi Railway Station.*

02 Dragon Well Tea Village

A classic Shànghǎi day trip is to the city of Hángzhōu, 'heaven on earth' according to Marco Polo. Veer away from the classically beautiful West Lake for the hills southwest, where the fine Dragon Well (*longjing*) green tea is grown. From Dragon Well Tea Village, walking trails lead into the plantations, and there are restaurants and tearooms where you can sample the brew fresh from the shrub. *8am–5.30pm; 1hr by high-speed train from Shànghǎi Hongqiao Railway Station.*

03 Zhūjiājiǎo

This charming canal-side village is filled with small temples and tiny bridges of the Ming and Qing dynasty architectural styles. Gondola-style boats ply the canals offering tours, but it's nice enough just to wander the alleyways, stopping for a cup of tea or a beer on one of the many terraces overlooking the water. *1hr by bus from Pu'an Rd bus station.*

04 Master Gao's Brewpub

One of the first craft brewers in China, Master Gao's taproom is in Nánjīng's 1912 nightlife district. Spread over two floors with an on-site microbrewery, the pub has 24 taps pouring Master Gao's own beers, plus international brews and pub grub – all the ingredients for a beer-themed breather. *6pm–midnight; 1hr 20min by high-speed train from Shànghǎi Hongqiao Railway Station to Nánjīng South.*

05 Mògànshān

Escape into the mountain mists at Mògànshān, a forest getaway developed as a summer retreat, popular with Europeans in the 19th century. Forest paths wind through bamboo groves, taking in a number of historic villas, including that of notorious Shànghǎi gangster Du Yuesheng, plus a house where Chairman Mao once took a nap. *1hr 30min by high-speed train from Shànghǎi Hongqiao station to Hángzhōu East, then train to Déqīng.*

——— TWO HOURS FROM ———

06 Xishantang Vegetarian Restaurant

In the old canal village of Tónglǐ, this Zen space combines fine dining and Buddhist pursuits, serving vegetarian meals in a meditative atmosphere. The interior is bedecked in soft wood, fabrics and scented candles. It doubles as a centre for Buddhist study, and hosts tea tastings, scroll painting and flower-arranging workshops. *129 Shangyuan Jie; 8am–8pm; 2hr by high-speed train from Shànghǎi Station to Sūzhōu, then metro to Tónglǐ, or by direct bus from Shànghǎi Bus Station.*

PǓTUÓSHĀN

In the Zhōushān Archipelago, Pǔtuóshān is one of China's four sacred Buddhist mountains and a celebrated isle, dedicated to the Goddess of Mercy, Guanyin. It's been popular with Shànghǎi weekenders for centuries, so aim for a midweek visit. Explore the island's sandy beaches, incense-filled temples and hidden meditation caves, then settle in for a vegetarian meal, or ferry to neighbouring Zhōushān to enjoy some of the day's catch at a seafood night market. It is 3–4hr by shuttle bus and speedboat from Luchao Port or 4hr 30min by bus from Shànghǎi South Long-Distance Bus Station.

Garden of the Master of the Nets, Sūzhōu.

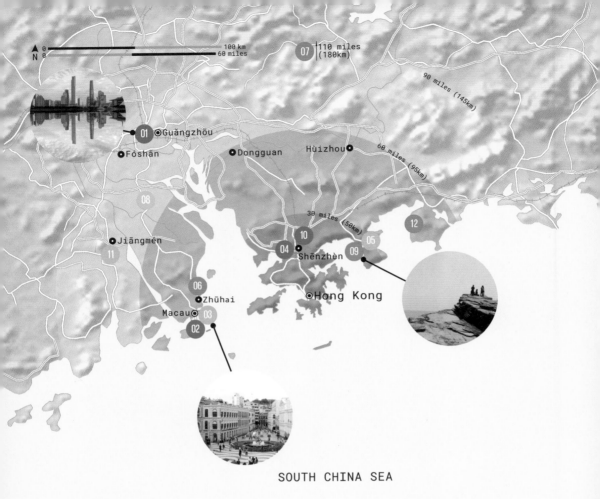

90 miles (145km)

01 ⊙Guǎngzhōu

⊙Fóshān

⊙Dongguan

Hùizhou⊙

60 miles (95km)

08

30 miles (50km)

12

⊙Jiāngmén

10
Shēnzhèn

05

04

09

11

06

⊙Zhūhai

Macau⊙ 03

02

⊙Hong Kong

SOUTH CHINA SEA

Life in one of the world's densest cities often calls for a head-clearing getaway.
A quick flight to somewhere tropical, overland (or overwater) is easy, but savvy
travellers could go luxe in Macau, rustic in Guangdong, or quirky in Shēnzhèn.

HONG KONG

ARTS & CULTURE ● HISTORY ● OUTDOORS ● FOOD & DRINK ● FESTIVALS & EVENTS ● MUSIC & FILM

——— ONE HOUR FROM ———

01 Guǎngzhōu
China's second-largest city is well off the tourist trail, which makes this sprawling mass of skyscrapers, canals and alleys a perfect pick for a weekend of no-expectations wandering. Try Shāmiàn Island, the 18th-century European trading post, for colonial mansions and shady lanes, Qingping Market for photogenic piles of dried medicinal herbs and seafoods, and Beijing Rd for teeming night markets. *50min by high-speed train from West Kowloon station.*

02 Cotai, Macau
Once a swampy marshland, Macau's Cotai area is now a fully fledged Asian Vegas. Casino sizes beggar belief: the 975,000 sq metre Venetian is Asia's biggest building, complete with an indoor faux Venice of canals and gondoliers. The Parisian boasts a half-scale Eiffel Tower, while the Galaxy has a wave pool, lazy river and fake beach... on the roof. Eat, gamble and revel in the kitschy magnificence of it all. *1hr by ferry from Macau Ferry Terminal.*

03 Macau
Casinos make you clammy? There's another side to Macau, one that's more about Portuguese culture and architecture than roulette. Start in Senado Square, where you'll find pastel colonial buildings and unique mosaic pavements. Proceed north to the 17th-century Ruins of the Church of St Paul, stopping for custardy egg tarts and sandy almond cookies. Then dine on spicy Macanese 'African chicken' amid the atmospheric alleyways of Taipa Village. *1hr by ferry from Macau Ferry Terminal.*

04 Shēnzhèn
Crossing the border for a weekend of spas and pedicures is a ritual for Hong Kong's *tàitai* (ladies of leisure). Follow their lead with a vigorous Chinese massage at the SLF International Spa Club, then gossip over fresh fruit in the lounge before high tea at the Ritz or the St Regis. Reserve half a day for knock-off purse shopping and cheap pedis at Lo Wu. *1hr 10min by metro from Central Station.*

05 Dapeng
Just east of the Shēnzhèn mega-city lies sleepy Dapeng Peninsula. It has several pleasant beaches and fresh seafood restaurants, but the big draw here is the walled city, built in 1394 to protect locals from the pirates that once prowled the South China Sea. Pass through the ramparts to wander stone alleyways lined with shops selling art and snacks, such as spicy noodles and sweet dumplings. *1hr 10min by taxi from Shēnzhèn border.*

06 Zhūhǎi
This Pearl River Delta town is best known for its hot springs, its beaches and its seafood. Make like a human hotpot at Zhuhai Imperial Hot Springs Resort, where you can soak yourself in everything from coffee to wine to milk to ginger, all set in manicured Japanese-style grounds. Then taxi across the peninsula to gorge on pick-your-own razor clams and garlic prawns at Wanzai Seafood Street. *1hr 10min by ferry from Macau Ferry Terminal.*

HIGH-SPEED RAIL

Hong Kong's new high-speed rail station courted controversy long before it opened in September 2018. Its joint passport checkpoint means part of the station is controlled by Běijīng, leading many pro-autonomy Hong Kongers to fear increasing mainland encroachment. Political issues aside, the station's bullet trains travel twice as fast as traditional trains to Guǎngdōng cities such as Guǎngzhōu, Shēnzhèn and Dongguan. The high-speed rail also connects Hong Kong with more distant mainland cities, such as Shànghǎi and Běijīng.

07 Danxiashan Geopark

Famed for its red sandstone pillars (including one notably phallic specimen; expect a queue for selfies), this Unesco geopark in northern Guǎngdōng province has hiking, cave temples and pavilions offering spectacular views across the emerald valley below. Those averse to hiking can tour the landscape via boat ride along an artificial lake. A futuristic high-speed train whisks you from Shēnzhèn to the park's gateway. *1hr 30min by train from the Shēnzhèn border.*

08 Shunde

This affluent district of Fóshān is considered by many to be the best place in Guǎngdōng province for Cantonese cuisine. Nibble the region's sweet steamed rice cakes at Sister Huan's Lunjiao Rice Cakes, which has been around for four generations. Taste velvety milk pudding and savoury pheasant rolls, both local specialities. For springy, shrimpy wonton noodles, head to any location of the beloved Yingji Noodle Shop. *1hr 30min by metro & high-speed rail from West Kowloon Station.*

—— TWO HOURS FROM ——

09 Tung Ping Chau

Although technically part of Hong Kong, this far-flung island certainly doesn't feel like it. Barely over 1 sq km in size, Tung Ping Chau is a tangle of banyan and palm jungle and decayed villages, ringed by rocky cliffs and wonderfully clean beaches. Hong Kong's most easterly point, its sunrises glow a psychedelic orange over the South China Sea, unimpeded by skyscrapers. Camp

05

© mary416 / Shutterstock

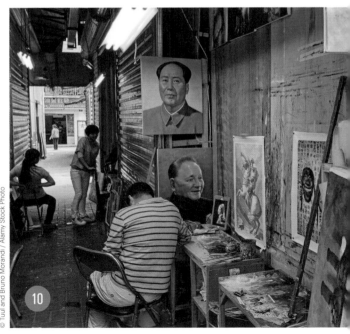

10

© Tuul and Bruno Morandi / Alamy Stock Photo

overnight to catch one in the morning. *1hr 40min by ferry from Ma Liu Shui Pier.*

10 Dafen Oil Painting Village, Shēnzhèn
In the market for a faux Van Gogh? On the outskirts of Shēnzhèn, Dafen Oil Painting Village cranks out an enormous proportion of the world's knock-offs, as well as original (if banal) works ideal for hotel lobbies. Watch artists paint in alleys and tiny studios, and even consider your own commission – many artists can reproduce your favourite photos in oils. *2hr 20min by metro from Central Station.*

11 Jiangmen
Along the banks of the Xijang River, this little-visited provincial city was once a treaty port, and still retains its historic colonial waterfront. Visit Ming dynasty-era Cha'an Temple, then cool off with a hike through the trees at Mount Guifeng National Forest Park, stopping to admire waterfalls and incense-fragrant shrines, and relax at one of the hot-springs resorts popular with domestic travellers. *2hr 30min by ferry from China Ferry Terminal.*

——— THREE HOURS FROM ———

12 Beaches of Eastern Guǎngdōng
For a sun and sand getaway that doesn't involve getting on a plane to Thailand, head to Daya Bay, home to several modest resort areas. Xunliao Bay, on the eastern side of Daya Bay, has some of the nicer hotels and will surprise you with its tropical-feeling beaches. To its south is China's only state-run sea turtle reserve, which is located on a stretch of gorgeously isolated sand. *3hr by bus and car.*

THE LOCAL'S VIEW

'People think of Hong Kong and think of skyscrapers, crowds and malls. But to get away from all that without actually leaving the city, head to Lantau Island. Take the ferry to remote Mui Wo and walk to even tinier Pui O beach, or catch the cable car from Tung Chung to the famous Big Buddha statue and then hike back. If you're feeling adventurous climb Sunset or Lantau Peak. Due to several long hiking trails and most of the islands protected in country parks, I can still find new beaches and villages even after years of living here in Hong Kong.'

Liz Jackson, professor at the University of Hong Kong

EUROPE

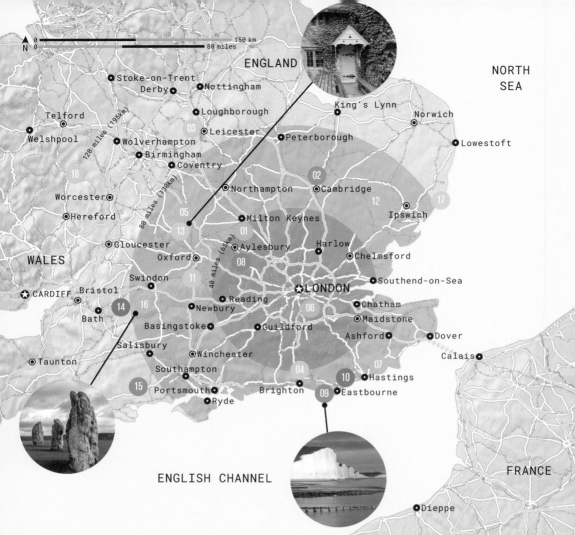

N
0 [scale bar] 150 km
0 [scale bar] 80 miles

ENGLAND

NORTH SEA

King's Lynn
Norwich
Lowestoft

Stoke-on-Trent
Derby
Nottingham
Loughborough
Telford
Leicester
Peterborough
Welshpool
03
Wolverhampton
Birmingham
Coventry
18
120 miles (195km)
Northampton
02
Cambridge
12
Ipswich
17
Worcester
Hereford
80 miles (130km)
05
Milton Keynes
13
01
Aylesbury
Harlow
Chelmsford
Gloucester
Oxford
08
Swindon
11
40 miles (65km)
Reading
LONDON
Southend-on-Sea
Chatham
CARDIFF
Bristol
14
16
Newbury
06
Maidstone
Bath
Basingstoke
Guildford
Ashford
Dover
Salisbury
Winchester
Calais
Taunton
Southampton
04
07
Portsmouth
15
Brighton
10
Hastings
09
Eastbourne
Ryde

WALES

ENGLISH CHANNEL

FRANCE

Dieppe

Explore medieval villages, hike along towering cliffs, nurse a pint in a historic country inn, feast on fine food and contemporary arts, or wander through villages of rose-clad cottages, all within easy reach of England's bustling capital city.

LONDON

ARTS & CULTURE ● HISTORY ● OUTDOORS ● FOOD & DRINK ● FESTIVALS & EVENTS ● MUSIC & FILM

——— ONE HOUR FROM ———

01 Bletchley Park

England's best-kept secret during WWII, Bletchley Park was a hive of brilliant minds who intercepted enemy messages and built machines to decrypt and interpret them. Among them was Alan Turing who cracked the infamous Enigma code. Volunteers bring the simple huts, near incomprehensible machines and everyday work of a codebreaker to life. *www.bletchleypark.org.uk; 40min by train from Euston.*

02 King's College Chapel, Cambridge

As close as you'll ever get to hearing a heavenly fanfare on earth, the celestial voices of the King's College Chapel choir and the magnificent setting, beneath mesmerising fan-vaulted ceilings and huge stained-glass windows, is quite simply mind-blowing. As soon as the gilded bat-wing organ lets fly a flurry of notes, you'll know you're in for something really special. *www.kings.cam.ac.uk/chapel/index.html; 50min by train from King's Cross.*

03 Golden Mile, Leicester

The heart of Leicester's Indian community for 40 years and renowned for its bazaar-like atmosphere, Belgrave Rd is a glorious tangle of brightly coloured saris, gold jewellery and fragrant spices. The dosas, idli and thalis in any of the local restaurants are some of the most authentic you'll find in Britain. Come for Diwali if you can – the massive celebrations here are internationally renowned. *1hr by train from St Pancras International.*

04 Brighton Festival

One of the most pioneering, participatory and innovative of arts festivals in Britain, Brighton rocks over three weeks each May with theatre, music, dance, film, art, circus, literature and family events. Experimental guest directors such as Anish Kapoor and Brian Eno lead the charge each year, ensuring the diverse programme – of what is now Britain's largest arts festival – never becomes banal or predictable. *www.brightonfestival.org; May; 1hr by train from Victoria.*

05 Broughton Castle

Undeniably grand yet somehow intimate, moated and fortified Broughton Castle is a romantic pile surrounded by gorgeous gardens. Its manageable scale and serene setting make it immediately enchanting. If Lord and Lady Sele are about, you may hear stories of the Fiennes' past adventures: it's been the family home since 1447, after all. *www.broughtoncastle.com; 1hr by train from Marylebone to Banbury.*

06 Down House, Kent

Pore over Darwin's notebooks in the house where he wrote *On the Origin of Species*. While you peruse the family home – where Darwin's hat, microscope and desk still sit – listen to Sir David Attenborough narrating an audio guide that explains the development of Darwin's groundbreaking theories and the controversy they caused. Outside, you can see the gardens and greenhouses where he conducted his experiments. *www.english-heritage.org.uk/visit/places/home-of-charles-darwin-down-house; 1hr 10min by car.*

<u>THE LOCAL'S VIEW</u>

'If you're looking for an urban escape in London, the Olympic Park offers plenty of green space, meadows and water gardens. Bring a book, cycle around and enjoy the evening sun. It's very quiet in many parts and the landscaping and planting is gorgeous. The park connects the River Lea in a really beautiful way and if you head north along the river from Hackney Marshes, it's a revelation. It's an area only known to locals, totally underused and you can walk all the way to Walthamstow Wetlands, Europe's biggest urban wetlands, which are just 15 minutes from central London.'

Lisa O'Carroll, journalist

© Valery Egorov / Shutterstock; © Justin Foulkes; © Martyn Ferry / Getty Images

07 Rye

A pocket of medieval magic, Rye's cobbled streets, lopsided, half-timbered buildings and wisteria-draped cottages seem ready-made for the Insta crowd. This former fishing, trading and smuggling port boomed as one of the Cinque Ports, five coastal towns with a royal charter to provide ships to the Crown. Rye now sits inland on a rocky outcrop several miles from the sea – apart from that, little has changed. *1hr 10min by train from St Pancras International.*

08 Wendover Woods

Leafy woods, rolling hills, sweeping views and utter tranquillity await in the scenic Chilterns. Largely undeveloped – apart from a treetop trail by the visitor centre and a kids' orienteering trail with a *Gruffalo* theme – the woods here remain quiet and undisturbed. An Iron Age hill fort hides in the undergrowth and views across the Vale of Aylesbury make this a great place to walk, run or ride. *www.forestry.gov. uk/wendoverwoods; 1hr 20min by car.*

09 White Cliffs, Seven Sisters

A thrilling, poignant and slowly disappearing wonder, the undulating white cliffs known as the Seven Sisters make a dramatic destination for a coastal walk. The cliffs are nibbled away by erosion and in places you can see buildings left precariously close to the edge. It's a 13-mile walk from Seaford to Eastbourne with shingle beaches, rock pooling and birdwatching to enjoy along the way. *1hr 30min by train from Victoria to Seaford.*

10 Towner Art Gallery, Eastbourne

One of the most exciting exhibition spaces in the southeast, this purpose-built gallery of contemporary art has a growing reputation for outstanding shows. Two floors showcase temporary exhibitions, while the 1st floor has rotating themed shows from the gallery's 5000-piece collection. Free tours offer behind-the-scenes access, while regular film screenings, talks and workshops offer activities for all. *www.townereastbourne. org.uk; 1hr 30min by train from Victoria.*

© Melinda Nagy / Shutterstock

11 Truck Festival

Small but perfectly formed, Truck is the antithesis of the big festival vibe, with a small crowd, low ticket prices, and profits from food takings going to local charities. Set in rural Oxfordshire on the last weekend in July, it has a broadly indie rock line up, but features everything from big-name headliners to new and emerging acts, heavy guitars, comedy, hip hop karaoke and cabaret. *www.truckfestival.com; late Jul; 1hr 30min by car.*

—— TWO HOURS FROM ——

12 Lavenham

Cosy up by the fire, sip a pint and let your photos go viral from Lavenham, possibly the prettiest – and wonkiest –

village in England. Hardly changed since the 15th century, the whole of this medieval wool town is delightfully wayward, its drunkenly lurching, half-timbered, ornately plastered and thatched buildings now converted into classy hotels and restaurants. It's about as romantic as the countryside gets. *1hr 50min by car.*

13 Falkland Arms, Great Tew

Squirrelled away in a tiny village of thatched cottages and wandering lanes, The Falkland Arms is a 16th-century boozer that makes the perfect place to end a Cotswold ramble. With flagstone floors, open fires, low beams and a collection of snuff, clay pipes and real ales behind the bar, it's about as authentic as it gets. *www.falklandarms. co.uk; 2hr by car.*

THE THAMES PATH

Take in some liquid history along the Thames Path, 184 miles of quiet waterside walking and one of the easiest ways to get out of the city. Passing through quaint villages and historical towns, water meadows, wetlands and fields of grazing cattle, it links you to places such as Henley, Oxford and Windsor. You can walk, run, pedal or paddle along the waterway, sit on a weir, go for a wild swim, hire a boat or simply soak up the scenery. Alternatively, wander as far as Hampton Court Palace, visit William Morris' house at Kelmscott or camp on an island at tranquil Pinkhill Lock.

© Victor Frankowski

14 Lacock Abbey

A 13th-century Augustinian nunnery turned stately home, Lacock Abbey is a deeply atmospheric place with a stunning Gothic entrance hall, medieval wall paintings and hushed cloisters. It was once home to William Henry Fox Talbot, a pioneering early photographer, and an exhibition traces his work and discoveries. Once you've had a gander, mooch around the nearby National Trust village for a taste of yesteryear. ***www.nationaltrust.org.uk/lacock-abbey-fox-talbot-museum-and-village; 2hr by car.***

15 Moors Valley, New Forest

A rare find, Moors Valley Country Park offers oodles of family-friendly outdoor adventure, yet all you pay for is parking. Along with two adventure playgrounds, there are mapped cycle trails, exercise stations and an innovative forest play trail, which will effortlessly persuade even the most reluctant explorers to venture further into the woods. You can also pay to hire a Segway, climb the treetops or ride a steam train. ***www.moors-valley.co.uk; 2hr by car.***

16 Avebury

Avoid the crowds at Stonehenge and make your way instead to Avebury, a stone circle so large it encompasses a village. You can walk right up to the massive stones here, clamber over earthworks and wonder whatever possessed Neolithic people to create this complex. The site is far bigger, older and quieter then Stonehenge and has more of a sense of mystery to it, too. ***www.nationaltrust.org.uk/avebury; 2hr by car.***

© simon evans / Alamy Stock Photo

© Dosfotos / Getty images

17 **Pinneys of Orford**

There's a subtle genius at work at Pinneys, a long-standing family-run oysterage and smokehouse in handsome Orford. Don't be fooled by the no-nonsense, canteen-like surroundings – the dishes have such delicate textures and depth of flavour that folk flock from far and wide to eat here. In the shop overlooking the quays you can pick up fish, poultry and meat cold-smoked over oak logs, plus plenty of pickles and chutneys to serve with them. *www.pinneysoforford.co.uk; 2hr 30min by car.*

THREE HOURS FROM

18 **Ludlow**

A medieval market town clustered around a Norman castle, set in a loop of the River Teme, Ludlow's wonky, half-timbered Tudor houses are the backdrop to a foodie enclave which, boasts top-notch restaurants, gastropubs, butchers, bakers, delis and cheesemongers. For peak indulgence, time your visit to coincide with the Ludlow Food Festival which is held over three days at the end of September.

3hr by train from Euston station.

PUBLIC TRANSPORT TIPS

Pore over any map of London and you'll find a surprising array of green space, both within the city and on its fringes, almost all of it easily accessible on the TfL (Transport for London) network. You'll pay a premium for a single journey ticket so avoid them at all times, even if you're just visiting. Use a contactless bank card, a travel card or an Oyster card instead. The TfL journey planner (www.tfl.gov.uk/plan-a-journey) will quickly find the best routes for you, and remember, children under 11 travel free on the underground, DLR and buses without a ticket.

THE UK'S MOST

With scenery like this, no wonder the Romans were so keen to invade Great Britain.

FASCINATING ROMAN SITES

Bignor Roman Villa

Do some research about Roman symbolism before you arrive and you'll get a deeper understanding of the mosaics at Bignor. Look out for Zeus the eagle abducting the beautiful Trojan soldier Ganymede, or the sadness of winter personified. *www.bignorromanvilla. co.uk; Pulborough, West Sussex; 10am-5pm Mar-Oct; £6.*

Caerleon Roman Fortress & Baths

Pretend you're a Roman soldier as you walk from frigidarium to tepidarium to gladiator match. Then look for clues about King Arthur who as, legend has it, held court here. *www.cadw.gov.wales; High St, Newport; 9.30am-5pm Mar-Oct; free.*

Fishbourne Palace

It turns out English gardens were a thing long before England even existed. See the UK's best example of a formal Roman garden before making a beeline inside to see the truly awesome 'Cupid riding a dolphin' mosaic. *www.sussexpast. co.uk; Roman Way, Fishbourne; 10am-5pm Mar-Oct, to 4pm Nov-Feb; £9.50.*

Guildhall Art Gallery & Roman Amphitheatre

Visiting this underground amphitheatre – only discovered in 1985 under an existing museum of 17th- and 18th-century art – feels like entering a secret and ancient world. *www.cityoflondon. gov.uk; Guildhall Yard, London; 10am-5pm Mon-Sat, noon-4pm Sun; free.*

Bath

Roman soldiers started coming here to 'take the waters' (as one-time resident Jane Austen would have said) almost 2000 years ago, and the Roman baths are some of the world's most perfectly preserved. *www.romanbaths. co.uk; Stall St, Bath; 9am-6pm, to 10pm 21 Jun-31 Aug; £16.50.*

Corinium Museum

Ever wondered why so many towns in the UK end in -cester, -chester or -caster? Cirencester's museum details the history of one of Britain's many Roman military forts (aka 'castra'). *www.coriniummuseum. org; Park St, Cirencester; 10am-5pm, to 4pm Nov-Mar; £5.60.*

York

Eboracum (York) was an important and wealthy walled city at the height of the Roman empire, and the Yorkshire Museum takes visitors through a local Roman citizen's domestic daily life. *www.yorkshiremuse-um.org.uk; Museum St, York; 10am-5pm; £7.50.*

Antonine Wall

A 39-mile earthen wall stretches from Scotland's Firth of Forth to the Firth of Clyde, built by Hadrian's successor, Antoninus Pius. A series of signposts along the way describe the region's Roman past. *www.antoninewall.org*

Chester

Sign up for an historic walking tour to explore the Roman walls, inscriptions and what was the largest Roman amphitheatre in Britain – it once held 8000 spectators for military drills, cockfights and, yes, even gladiator contests. *www.english-heritage. org.uk/visit/places/ chester-roman-amphi-theatre; Little St John St, Chester; dawn-dusk; free.*

Vindolanda

Vindolanda was already a popular Roman site along Hadrian's Wall when, in 2017, archaeologists discovered 2000 buried Roman artefacts, including still-shiny Roman cavalry swords and woven cloth. Bonus: two-week volunteer archaeologist opportunities. *www.vindolanda.com; Bardon Mill, Hexham; 10am-6pm, closed 1 Nov-10 Feb; £7.50.*

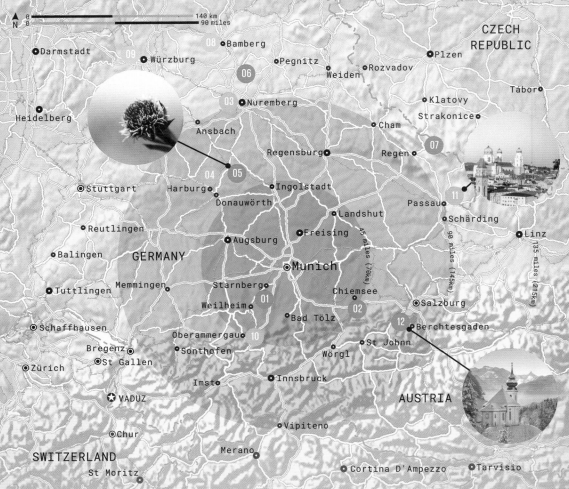

N

140 km
90 miles

Darmstadt

09 Würzburg

08 Bamberg

06

Pegnitz

Weiden

Rozvadov

CZECH
REPUBLIC

Plzen

Tábor

Heidelberg

03 Nuremberg

Ansbach

Cham

Klatovy

Strakonice

Stuttgart

Harburg

04 05

Donauwörth

Regensburg

Ingolstadt

Regen

07

Passau

11

Schärding

Linz

Reutlingen

Augsburg

Freising

Landshut

45 miles (70km)

90 miles (145km)

135 miles (215km)

Balingen

GERMANY

Starnberg

Munich

01

Chiemsee

Salzburg

Tuttlingen

Memmingen

Weilheim

Bad Tölz

02

Berchtesgaden

12

Schaffhausen

Oberammergau

10

St Johnn

Bregenz

Sonthofen

Wörgl

AUSTRIA

Zürich

St Gallen

VADUZ

Imst

Innsbruck

Chur

Vipiteno

SWITZERLAND

St Moritz

Merano

Cortina D'Ampezzo

Tarvisio

ITALY

From Munich, on a clear day, you can just make out Germany's sliver of the Alps looming invitingly on the horizon. Other Bavarian getaways include beer festivals and canoeing, Roman ruins and mysterious forests full of berries and mushrooms.

MUNICH

● ARTS & CULTURE   ● HISTORY   ● OUTDOORS   ● FOOD & DRINK   ● FESTIVALS & EVENTS   ● MUSIC & FILM

——— ONE HOUR FROM ———

01 Fünf-Seen-Land

When Munich folk grow tired of the urban hullaballoo, they head south to the Fünf-Seen-Land (Five Lakes Area). Starnberg by the Starnberger See is the unofficial capital of the region. Set out from here on the 5km hike to Berg where King Ludwig II drowned in uncertain circumstances. Next to the Ammersee is Andechs, with its famous monastery where monks produce one of Bavaria's best beers. *30min by S-Bahn.*

02 Chiemsee

The highlight of the Chiemsee, Bavaria's largest lake, is an island just offshore occupied by one of the state's most famous chateaux – Ludwig II's Herrenchiemsee. Reached by ferry from lakeside Prien, this schloss was inspired by Versailles and features opulent interiors, which Ludwig II never actually saw completed. The Chiemsee has other islands to discover, as well as water sports galore. *1hr by train.*

03 Nuremberg

Hop aboard a train for the short trip to Bavaria's second city. Nuremberg is an engaging place whose rich and troubled history is uncovered in medieval castles, the unsurpassed German National Museum and Deutsche Bahn Museum, and the Nuremberg Trials memorial. In between sights, stop off for some of the city's famous bratwurst, finger-sized sausages, best accompanied by a dark regional lager. *1hr by train.*

04 Harburg Castle

One of Bavaria's most striking castles, Schloss Harburg rises dramatically above the River Wörnitz. Perfectly preserved in its medieval splendour, Harburg is a popular stop on the so-called Romantic Road, Germany's most popular tourist route. Best of all, if you don't fancy the train ride back to Munich, part of the fortress is now a hotel, surely one of the region's most distinctive places to stay. *www.burg-harburg.de; 1hr 30min by train.*

05 Altmühltal Naturpark

If you feel the need to work off all that beer and bratwurst, try canoeing the Altmühl River Valley, an easy-going 100km of picturesque landscapes lined up along the slow-moving waters of this Danube tributary. Canoe hire, food and accommodation options are centred around the Italianate town of Eichstätt, a worthwhile destination in its own right, featuring handsome 17th-century architecture, cobbled streets and leafy piazzas. *www.naturpark-altmuehltal.de; 1hr 30min by train.*

———TWO HOURS FROM———

06 Franconian Switzerland

For a bucolic retreat, head to the Franconian Switzerland, an area known for its ancient hills, babbling brooks, tiny breweries and weird-and-wunderbar rock formations. Almost every village in these parts runs a brewery producing Landbier, a regional draught that's usually only available locally. The area is also known for its folksy, forest food – burn off all that venison and rabbit with a hike or a spot of rock climbing. *2hr by car.*

DISCOVERING BAVARIAN BEER

It's no secret that Bavaria produces some of the world's best beer. The quality of local suds is guaranteed by the 1516 Reinheitsgebot (Purity Law), which states that beer may only contain three ingredients – hops, barley and water. Munich's big six breweries are world famous, as is the Oktoberfest, but beyond the capital there's a galaxy of foaming tankards to discover. The town of Straubing holds one of the biggest beer festivals in Germany, Erding is renowned for its wheat beer, Freising has what is possibly the world's oldest brewery, while northern Bavaria is known for its microbrewed Landbier.

07 Bavarian Forest National Park

Running down the Czech border, this national park is a real antidote to urban Bavaria. Mirroring similar wooded hills on the Czech side of the former Iron Curtain, this small protected landscape abounds in unspoilt forests, gurgling streams, forgotten border villages and lots of wildlife. The area is also well known for its glass – the town of Frauenau has Germany's best glass museum. *www.nationalpark-bayerischer-wald. bayern.de; 2hr by car.*

08 Bamberg

Beautiful Bamberg is synonymous with its astonishing architecture, but another appealing aspect of this delightful town is its beer. The award-winning BierSchmecker tour run by the tourist office takes in an unbelievable eleven breweries located within the city limits, some of which produce Bamberg's unusual smoke beer, Rauchbier. Of course, the tour includes plenty of tasting and there's no guarantee you'll remember the latter stages. *www.en.bamberg.info/ bierschmeckertour; 2hr by train.*

09 Würzburg

Bavaria may be more synonymous with hops than grapes, but Würzburg, in the far north of the state, bucks the trend. On alighting at the station, oenophiles will be reassured to see vineyards providing a verdant backdrop to this lively city. The best place to head for some local whites is the Bürgerspital, where they'll even serve you soup made from wine. *www.buergerspital.de; 2hr by train.*

10 Oberammergau

Pretty alpine Oberammergau is known for its passion play, the story of Jesus' last days, which local people perform every ten years. When you've toured the Passion Theatre, take the 9km hike to Linderhof Castle, the least-visited of Ludwig II's fairy-tale castles,. Alternatively, nearby Ettal Monastery produces its own herbal liqueur – what is it with these monks and booze? *2hr by train.*

03

11 Passau

At the confluence of three rivers, the Inn, Danube and Ilz, the town of Passau is a real treat for architecture buffs, with everything from Roman ruins to bold 21st-century statements to admire. The churches, palaces and townhouses seem to be piled one on top of the other, crammed onto a promontory that dips down to the three rivers, where waters create a murky tricolour as they meet. **2hr 20min by train.**

12 Berchtesgadener Land

Some particularly spectacular Alpine scenery can be found in Bavaria's far south. The Berchtesgadener Land was a favourite with Hitler et al. and the region has some fascinating dark tourism, but the real reason to come is for an electric boat trip along the Königssee, an alpine lake surrounded by towering limestone peaks. There are also hikes galore affording impressive holiday-brochure-style mountain vistas. **2hr 30min by train.**

MOVING MUNICH

Munich has one of Europe's best public transport systems and one element of it, the S-Bahn, is ideal for getting you out of the city. For destinations further afield, the Hauptbahnhof is served by an alphabet spaghetti of train services including Deutsche Bahn (DB) and BOB, the latter heading south into the Alps. Often, special tickets are available that combine a train ride with an activity, such as skiing or visiting a major attraction. These are offered by DB and can save you a lot of euros.

08

Courtesy of Schlenkerla Holzfassausschank

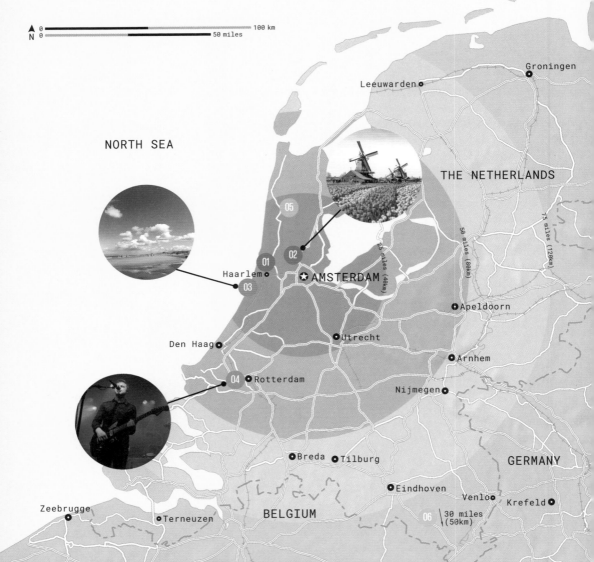

NORTH SEA

THE NETHERLANDS

Groningen

Leeuwarden

05

02

01

Haarlem

03

★ AMSTERDAM

25 miles (40km)

50 miles (80km)

75 miles (120km)

Apeldoorn

Den Haag

Utrecht

Arnhem

04 Rotterdam

Nijmegen

GERMANY

Breda Tilburg

Eindhoven

Venlo Krefeld

Zeebrugge

BELGIUM

06 30 miles (50km)

Terneuzen

0 100 km

0 50 miles

N

All of the Netherlands is a getaway from Amsterdam. Little in this country is over three hours away by frequent train. In no time at all you can enjoy endless beaches, pulsing clubs, old masterpieces, iconic festivals and, yes, windmills.

AMSTERDAM

● ARTS & CULTURE ● HISTORY ● OUTDOORS ● FOOD & DRINK ● FESTIVALS & EVENTS ● MUSIC & FILM

——— ONE HOUR FROM ———

01 Frans Hals Museum, Haarlem
This classic Dutch city of cobbled streets, historic buildings, grand churches, cosy bars, fine cafes and canals is also host to an exceptional art museum. The Frans Hals Museum is a treasure chest of Dutch Masters. Located in the poorhouse where Hals spent his final years, the collection focuses on the 17th-century Haarlem School, including Hals' Civic Guard series. *www.franshalsmuseum.nl; 20min by train.*

02 Zaanse Schans
People come to Zaanse Schans for an hour but find they stay for several. On the banks of the Zaan river, this open-air museum is a fabulous place to see windmills in action. Visitors can explore the six operating windmills at their leisure, and see how these devices are a combination of sailing ship and Rube Goldberg contraption. *www.dezaanseschans.nl; most windmills 9am-5pm Apr-Oct, hours vary Nov-Mar; 20min by train.*

03 Zandvort
Just west of Haarlem lies Zandvoort, a popular beach town. While the town isn't a stunner, the seemingly endless, flat, dune-backed beaches are. Walk just a little north or south to escape crowds. About 3km north is Bloemendaal aan Zee, a wonderfully undeveloped spot with uninterrupted beaches. Also nearby is Zuid Kennemerland National Park, which has dune hiking trails and plenty of wildflowers. *40min by train.*

04 Rotterdam
Music dominates the city's entertainment scene and draws partygoers from across Europe. Many live-music venues resist being confined to one genre, programming everything from jazz to rock, reggae to R&B. With the tunes come expertly made cocktails and delicious microbrews. The clubs are some of the best west of Berlin – don't miss huge and thumping Massilo and DJ-favourite Noah. *www.rotterdam.info; 40min by train.*

05 Museum BroekerVeiling, Broek op Langedijk
Waterlogged north Holland once had 15,000 tiny farms, each one an island, whose farmers tended their crops by rowing boat. You can tour the auction house where farmers used to arrive with boatloads of produce then waited – afloat– until they could paddle through an auction room where buyers would bid on their wares. Interactive exhibits and a mock auction recreate this unique life. *www.broekerveiling.nl; 10am-5pm Jul & Aug, closed Mon Sep-Jun; 1hr 30min by train & bus.*

——— TWO HOURS FROM ———

06 Maastricht Carnaval
Everything stops for Maastricht's Carnaval, one of Europe's wildest. The orgy of partying and carousing begins the Friday before Shrove Tuesday and carries on until the last person collapses sometime on Wednesday. Highlights include the hoisting of the clog-footed *Moosweif* ('Cabbage Woman'). *www.gemeentemaastricht.nl; Feb or Mar; 2hr 30min by train.*

THE LOCAL'S VIEW

'I don't like driving. I prefer to drink. When I need to leave Amsterdam, I take the train. It runs four to six times an hour and I get to Rotterdam in 30 minutes by high-speed Intercity Direct. Fun Antwerp is one and a quarter hours away, boring Brussels is two hours and delightful Paris is only three and a bit. I'm embarrassed to say I never have been in the Chunnel, but at 45 minutes, the plane to London is convenient, cheap and frequent. You can go on a whim.'

Andrew Moskos, comedian & founder, Boom Chicago

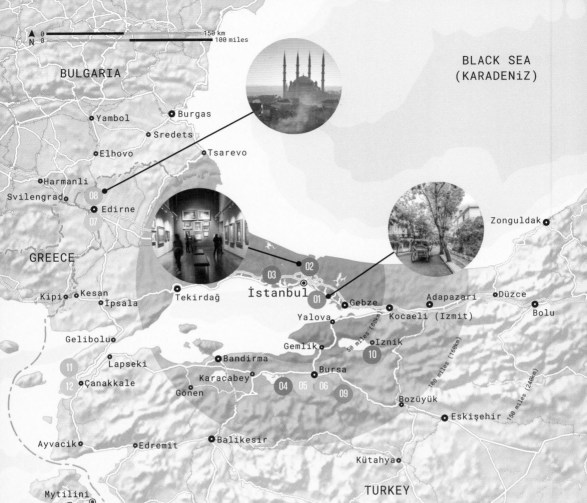

BULGARIA

• Yambol
• Burgas
• Sredets
• Elhovo
• Tsarevo
• Harmanli
Svilengrad•
08
07
• Edirne

GREECE

Kipi• •Kesan
•İpsala
• Tekirdağ
03
02
İstanbul
01
• Gebze
Yalova•
Gelibolu•
• Adapazari
Kocaeli (Izmit)
Düzce
Bolu•

BLACK SEA
(KARADENiZ)

Zonguldak•

Lapseki
11
12
•Çanakkale
• Karacabey
Gönen•
• Bandirma
Gemlik•
04 **05** **06**
Bursa
09
İznik
10
50 miles (80km)
100 miles (160km)
150 miles (240km)
Bozüyük•
• Eskişehir

Ayvacik•
• Edremit
• Balikesir
Kütahya•

TURKEY

Mytilini
Town©

0   150 km
0   100 miles
N

Within easy reach of Turkey's greatest city are surprising institutions and events,
evidence of the culinary and architectural legacy of the Ottoman Empire, and the
poignant memories of one of the 20th-century's most tragic battles.

İSTANBUL

⬤ ARTS & CULTURE ⬤ HISTORY ⬤ OUTDOORS ⬤ FOOD & DRINK ⬤ FESTIVALS & EVENTS ⬤ MUSIC & FILM

─── ONE HOUR FROM ───

01 Princes Islands
A relaxing haven from the incessant energy of İstanbul, the Princes Islands are around 20km southeast of the city. Five of the nine islands are inhabited and most popular for visitors are Büyükada and Heybeliada. Catch a ferry from İstanbul's Eminönü docks and combine oceanfront dining with walking in shaded pine groves and swimming in tiny coves. Try to travel on a quieter weekday. *1hr by ferry from Eminönü.*

02 Sakıp Sabancı - Museum
It's a win-win for visitors to İstanbul's Sakıp Sabancı Museum. In the wealthy waterfront neighbourhood of Emirgan, the museum is renowned for its superb permanent collection of Ottoman manuscripts and calligraphy, but temporary exhibitions are also hugely popular. Past shows have included works from such art superstars as Monet, Miro and Dalí, and important historical displays featuring the influence of Genghis Khan. *sakipsabancimuzesi.org; 1hr by bus from Taksim.*

03 Sancak Mosque
In a country endowed with beautiful heritage mosques, the Sancaklar Camii (Sancak Mosque) is a contemporary counterpoint to the work of past Ottoman architectural luminaries such as Sinan. Built between 2011 and 2012, the mosque is constructed in soothing grey stone, and designed to subtly conform to the contours of a natural amphitheatre. The design

continues into the subterranean interior, starkly beautiful, but still promoting deep reverence and contemplation. *1hr by car.*

─── TWO HOURS FROM ───

04 Karagöz Museum
Originally developed as a portable form of entertainment by travelling nomads from Central Asia, the art of *karagöz* (shadow puppetry) reached its zenith in Bursa during the Ottoman Empire. This museum in Bursa features puppets from around the former Ottoman Empire. Most Wednesday mornings at 11am, a thoroughly entertaining puppet show takes place. Prepare to cheer and boo in all the right places. *hometurkey.com/en/attractions/karagoz; 2hr by bus and ferry from Esenler.*

05 Kebapçı İskender
Foodie travellers will find Iskender kebabs all over Turkey, but the original version of the meaty treat, reputedly first created in 1867, is available at this much-loved heritage restaurant in the city of Bursa. Tender slices of grilled lamb are combined with a rich tomato sauce and small squares of pitta bread, and then lavishly topped with melted butter and yoghurt. *iskender.com.tr; 2hr by bus and ferry from Esenler.*

06 Yeşil Türbe
The birthplace of the Ottoman Empire, Bursa hosts many important architectural statements, but amid a cityscape of imposing mosques, markets and museums, one of Bursa's smaller

© Nikolay Dimitrov / Shutterstock; © Nejdet Duzen / Shutterstock; © Tim E White / Alamy Stock Photo

structures is its most poignant and beautiful. Framed by a cypress-trimmed park, the compact Yeşil Türbe ('Green Tomb') dates from the 15th century, and the mausoleum of Sultan Mehmed I is simultaneously simple and sublime. *8am-noon & 1-5pm; 2hr by bus and ferry from Esenler.*

——— THREE HOURS FROM ———

07 Kırkpınar Oil Wrestling Festival

Known locally as the Tarihi Kırkpınar Yağlı Güreş Festivali, this annual extravaganza of muscular men clad in tight leather shorts draws attendees from around the planet. Smeared liberally with olive oil, competitors attempt to force a submission from their opponent. In Turkey, kırkpınar is regarded as a very serious and also very macho sport, so no sniggering at the back, please. *kirkpinar.org; Jun or Jul; 3hr by bus from Esenler.*

08 Selimiye Mosque

Mimar Sinan (1497–1588) is regarded as the Ottoman Empire's finest architect, and this mosque built on Edirne's highest point is his greatest work. Four towering minarets, reaching a further 70m, are visible across the city, and the structure's scale is balanced by delicate touches like marble fountains and colourful interior calligraphy. Architectural fans will find Edirne has many other sublime and impressive buildings. *3hr by bus from Esenler.*

09 Mt Uludağ

Mt Uludağ near Bursa is Turkey's premier ski resort, but the sprawling alpine leviathan is also a popular destination for İstanbul residents during spring and summer. The Uludağ Teleferik, one of the world's longest gondolas, takes more than 20 minutes to ascend the mountain. En route to the gondola's terminus, enjoy rustic teahouses, barbecue restaurants and wooded walking trails at the Sarıalan stop. *3hr by car.*

10 İznik Foundation

İznik's tradition of tile-making began in the early 16th century, and the distinctive blue and white tiles originally crafted by artisans from the Persian city of Tabriz were used for the Ottoman Empire's most important buildings. In a new century, the İznik Foundation is resurrecting the lakeside town's tile-making heritage, and travellers can visit its fascinating open workshop. *iznik.com/en; 3hr by ferry & bus via Yalova.*

11 Gallipoli Historical National Park

This forested peninsula is dotted with around 60 war cemeteries, the tragic toll of

© zkan ulucam / Shutterstock

WWI battles between Allied and Turkish forces throughout 1915. Most poignant for New Zealanders is Anzac Cove – a compact bay washed by the cobalt waters of the Aegean – while the battle for Lone Pine is writ large in Australian history. Gallipoli is also becoming a popular destination for contemporary Turkish travellers. *http://nationalparksofturkey.com/ gelibolu-gallipoli-peninsula-historical-park; 3hr 30min by bus from Selimpasa.*

12 Suvla Winery

A surprisingly cosmopolitan destination near the sleepy town of Eceabat, the Suvla Winery combines wines made from 60 hectares of organic vineyards with an excellent terrace restaurant. The most highly regarded varietals are Suvla's zingy sauvignon blanc and spicy merlot, and local grapes Kınalı Yapıncak and Karasakız are unique to the surrounding region of Thrace. *suvla.com.tr; 4hr by car.*

© ozkan ulucam / Shutterstock

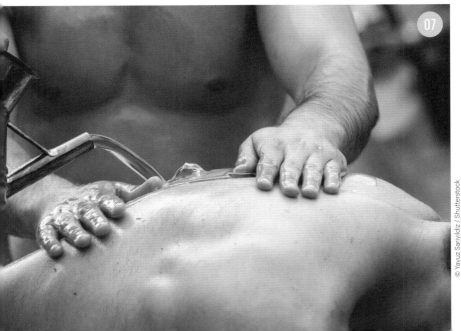

07

© Yavuz Sarıyıldız / Shutterstock

EXPLORING THE BOSPHORUS BY FERRY

İstanbul offers many ways to explore the Bosphorus Strait. From the Eminönü docks, catch a ferry from Europe to Asia, to the markets and restaurants of up and coming Kadıköy (20min). At the northern end of the Bosphorus, near the Black Sea, the fishing village of Anadolu Kavağı (1hr30min) is a popular destination for a leisurely seafood lunch. Two medieval Ottoman fortresses steeped in history are essential attractions en route. Built in 1453, Rumeli Hisarı is now a museum and is also used for concerts, while on the opposite shore, Anadolu Hisarı dates back to 1394.

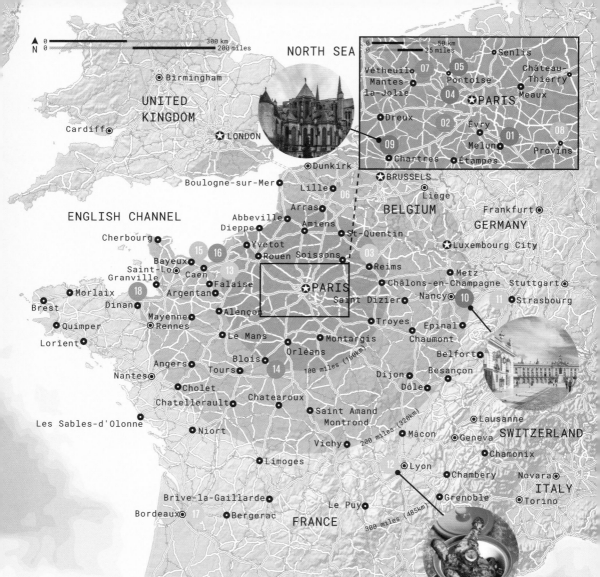

Inset map (upper right)

North Sea

07 · Vétheuil · Mantes-la-Jolie · 05 Pontoise · 04 · Château-Thierry · Senlis · Meaux · PARIS

Dreux · 02 · Évry · Melun · 01 · 08 Provins

09 · Chartres · Étampes

Main map

0 300 km
0 200 miles

NORTH SEA

Birmingham

UNITED KINGDOM

Cardiff

LONDON

Dunkirk

Boulogne-sur-Mer

BRUSSELS

ENGLISH CHANNEL

Lille

Arras 06

BELGIUM

Liège

Frankfurt

GERMANY

Abbeville · Amiens · St-Quentin

Dieppe

Cherbourg

Yvetot · 15 · 16 · Rouen · Soissons · 03 · Reims · Metz · Stuttgart

Bayeux · 13

Saint-Lo · Caen · Châlons-en-Champagne · Strasbourg

Granville · Falaise · PARIS · Nancy · 10 · 11

Morlaix · 18 · Argentan · Saint Dizier

Brest · Dinan · Alençon · Troyes · Epinal · Chaumont

Quimper · Mayenne · Rennes · Le Mans · Montargis · Dijon · Belfort · Besançon

Lorient · Angers · Blois · Orléans · 100 miles (166km) · Dôle

Nantes · Tours · 14 · 200 miles (328km) · Lausanne

Cholet · Chatearoux · Mâcon · Geneva · SWITZERLAND

Chatellerault · Saint Amand Montrond · Vichy · Chamonix

Les Sables-d'Olonne · Niort · 12 · Lyon · Chambery · Novara · ITALY

Limoges · Grenoble · Torino

Brive-la-Gaillarde · 17 · Bergerac · Le Puy · 300 miles (485km)

Bordeaux · FRANCE

PARIS

Paris' north of centre location, and France's great rail system, make for myriad escapes: pretty wine regions, Unesco-listed architecture, natural wonders, seaside towns and gourmet getaways.

● ARTS & CULTURE ● HISTORY ● OUTDOORS ● FOOD & DRINK ● FESTIVALS & EVENTS ● MUSIC & FILM

—— ONE HOUR FROM ——

01 Château de Vaux-le-Vicomte
This is the place to experience the unbridled splendour and wealth of 17th-century French aristocracy, minus the hordes. Famously inspired by his 1661 visit here, Louis XIV hired the same architect and landscape architect to design the opulent palace at Versailles. (Some say the Vaux-le-Vicomte gardens are Le Nôtre's real masterpiece.) At Christmas and summertime, the chateau and grounds are illuminated by candlelight.
www.vaux-le-vicomte.com; 40min by train from Gare de l'Est to Verneuil l'Étang.

02 Chevreuse Valley
The Upper Chevreuse Valley Regional Nature Park extends over 250 sq km, with walking trails criss-crossing its forests, pastures, lakes and rivers. Plan your randonnée to take in some of the park's heritage treasures, such as Château Dampierre with gardens by Le Nôtre (of Versailles fame) and Château Breteuil, livelier than most with its waxworks and hedge maze. *50min by train to Saint-Rémy-lès-Chevreuse from Gare du Nord.*

03 Champagne country
The unofficial capital of the Champagne region, Reims is a town with a sparkle in its step. Take a peek at the impressive cathedral (French kings were crowned here for over 1000 years), then hit the Champagne houses: Veuve Clicquot, Tattinger, Pommery and Mumm all offer cellar tours and tastings. Discover smaller winemakers and gorgeous vine-and-village scenery along parts of the 400km Route du Champagne. *1hr by train from Gare de l'Est.*

04 Château de Monte-Cristo
When planning his new home, Alexandre Dumas, author of *The Count of Monte Cristo*, said he wanted 'a Renaissance château, faced by a Gothic pavilion, surrounded by an English garden' – and that's what he got. Today the mini-palace is a place of literary pilgrimage. Dumas called the Gothic pavilion Château d'If, after a location in the novel that gives the whole place its name.
www.chateau-monte-cristo.com; 1hr by train from Gare de Lyon.

05 Auvers-sur-Oise
In the short time he spent in this pretty village on the Oise River, Vincent van Gogh completed almost 90 paintings, some of the most famous of his career. Wander the town to see the places he painted, visit his room at the auberge, and walk up to the remote little cemetery where he and his brother lie buried side by side. *1hr by train from Gare du Nord.*

06 Lille
Vieux Lille, France's Flemish heart, is a 17th-century town that feels more like a bite-sized Brussels than a French regional centre, with its exuberant facades of bright colours or patterned red brick, steeply gabled roofs and elaborate parapets. Get lost roaming the winding cobbled streets; as you're in France's beer capital, slip into a cosy bar to sample the local Belgian-style brews. *1hr by train from Gare du Nord.*

ESCAPE TO THE SOUTH OF FRANCE

France's TGV (train à grande vitesse; very fast train) system means that even the sunny south is within reach for a weekend break. In just over three hours you're in Marseille, the grand-and-gritty hub of southern France that trend-watchers are calling 'Berlin-on-sea'. Its wide blue skies, vibrant Old Port and proximity to the cliffs and bays of the stunning Calanques have long attracted visitors; these days, Marseille's swanky new museums, exciting dining and buzzing cultural scene cram the weekend must-do list.

 © Factofoto / Alamy Stock Photo

07 The Seine

On the western edge of the Parc naturel régional du Vexin français, the Seine crooks its arm before flowing on into Normandy, cradling a secret little nature reserve. Well, not completely secret: this spot was a favourite of Impressionist painters, including Monet. Take a canoe out on the gentle waters to experience a natural world largely unchanged since the artists set their easels on these tranquil banks. ***www.pnr-vexin-francais.fr; 1hr 20min by car to Vétheuil.***

08 Provins

Unesco-listed as one of the most intact medieval walled towns in France, Provins was a venue for the great trade fairs that linked northern Europe with the Mediterranean. The town's hilltop tower and 5km of ramparts make a forbidding display of 12th-century military architecture. Come in June for France's largest medieval festival to see jousting, jesting and other forms of costumed historic revelry. ***www.provins-medieval.com; 1hr 30min by train from Gare de l'Est.***

────── TWO HOURS FROM ──────

09 Chartres

Some declare it the most beautiful cathedral in France. It's said to be the finest example of French Gothic architecture. One thing is for sure: the stained-glass windows at Chartres' Cathédrale Notre Dame are a true masterpiece. Created in the 13th century, they depict the labours of local tradespeople as well as traditional biblical scenes. Look out for the famously vivid, unique blue – Chartres blue. ***www.cathedrale-chartres.org; 1hr 40min by train from Gare d'Austerlitz.***

10 Nancy

When Prussia annexed Alsace-Lorraine in 1871, Nancy became the capital of Eastern France, and home to thousands of refugees. The École de Nancy – artists and architects working in the Art Nouveau style – created homes and decorative arts to meet the refined tastes of a wealthy population. Immerse yourself in their Art Nouveau interiors at Musée de l'École de Nancy and Villa Majorelle. ***www.ecole-de-nancy.com; 1hr 40min by train from Gare de l'Est.***

11 Alsace Christmas markets

Strasbourg, in France's German corner, holds the country's biggest and most sparkling *marché de Noël*. Around the

 11

medieval old town, bedecked with twinkling light displays, clusters of little wooden chalets offer Alsatian crafts, wintry specialities like gingerbread, sausages and sauerkraut (choucroute here) and of course, warming *vin chaud* (mulled wine). Nearby, colourful half-timbered Colmar becomes a twinkling fairy-tale village.

1hr 50min by train from Gare de l'Est.

12 Lyon

When the French want to eat really well, they come to Lyon. The city's famed *bouchons* –convivial family-owned places serving up hearty local cuisine, such as Lyonnaise sausage, duck pâté, coq au vin – are famed for fun times and belt-busting portions. Don't miss a visit to covered market Les Halles de Lyon Paul Bocuse, overflowing with local specialities: cheese, quenelles (creamed fish dumplings) and countless ways with pigs' innards.

2hr by train from Gare de Lyon.

13 Normandy

If the little town of Pont-L'Évêque sounds familiar, that may be because you've eaten it. Its cheese is the oldest of Normandy's four AOC fromages – the others are Camembert, Neufchâtel and Livarot. Normandy is also famous for its apples, and in the small area of Calvados they're distilled into apple brandy. Visit the 17th-century half-timbered buildings of Maison Drouin to taste its hyper-local heritage drop.

2hr by train from Saint-Lazare.

14 Loire Valley

The gorgeous natural setting of the Loire River, magnificent chateaux, a dense network of well-signposted (and mostly flat) routes, plus fresh and fruity wines – the Loire is heaven on a bike. Set a relaxed pace along quiet roads through AOC vineyards, making regular pit stops for wine tastings, or to enjoy a home-cooked meal at a family-run bistro.

2hr by train to Blois from Saint-Lazare.

AMIENS' GIANT FLEA MARKET

Once a year, the streets of Amiens (1hr 20min by train) are invaded by an army of vintage aficionados and *Antiques Roadshow* hopefuls for one of France's biggest flea markets. Brocantes are a national pastime, and each October more than 2000 stallholders lay out their wares for this shopping frenzy that takes over the town. You're sure to find a must-have amongst the heritage treasures and retro knick-knacks. Beyond the bargains, there's an immense Unesco-listed Gothic cathedral and an excellent fine arts museum to explore.

© Michal Szymanski / Shutterstock

15 The Bayeux Tapestry

History's first comic book, the curved 70m of the Bayeux Tapestry sit behind glass in a dark room; visitors traverse its length slowly, like the faithful before a religious relic. Tracing the bloody tale of the Norman Conquest of 1066, the sense of rising drama as the Battle of Hastings draws near is intense; it's hard to believe that stitches in woollen thread can be so moving. ***www.bayeuxmuseum.com; 2hr 20min by train from Saint-Lazare.***

16 Deauville

Parisians' favourite beach getaway, Deauville exudes belle-époque glamour with its chic modern boutiques and near-constant buzz of film festivals, horse races, yachting regattas, jazz festivals and vintage car rallies. Hire a famously gaudy parasol and laze on the white sand, stroll the 1920s boardwalk to admire the grand seafront villas, and tuck into a giant pot of moules marinières at iconic brasserie Les Vapeurs. ***2hr 30min by train from Saint-Lazare.***

—— THREE HOURS FROM ——

17 **Wine in Bordeaux**

Hop the fast train for lightning lessons on wine history and theory at the stunning Cité du Vin, an ambitious multi-sensory wine museum. Then wander Bordeaux's elegant avenues and medieval lanes to taste your way through the region's drops in the city's charming wine bars. Hit the swanky bar at the École du Vin for a delicious final exam. *2–3hr by train from Montparnasse.*

18 **Saint-Malo**

Saint-Malo is a storybook walled town on the Brittany coast with a beautiful beach, atmospheric Breton-style pubs and a crêperie on every cobblestoned street. Come August, it's the site of one of France's best music festivals, La Route du Rock. The main stage is an 18th-century fort outside town; another is set right on the beach, with the city's stone ramparts providing a stunning backdrop. *3hr by train from Gare du Nord.*

THE LOCALS' VIEW

'Villers-sur-Mer: big beaches, a lovely little town centre, a casino (for those who like that kind of thing), long walks in the middle of a protected nature zone and the Paléospace science museum.' Gaëlle C-G

'Le Quesnoy, a fortified village with magnificent ponds and the Forest of Mormal – a peaceful haven just two hours away by train.' Julie L

'Baie de Somme is one of the most beautiful getaways I've been lucky enough to visit. The sea, the peace, the seal reserve – it's a real pleasure either alone or with someone special.' *Noria L*

18

© AGaeta / Getty Images

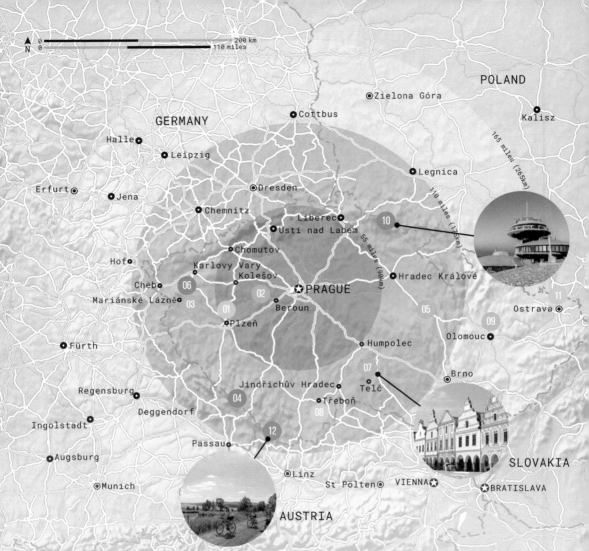

N

| 0 | 200 km |
| 0 | 110 miles |

POLAND

GERMANY

Zielona Góra

Cottbus

Kalisz

Halle

Leipzig

Legnica

165 miles (265km)

Erfurt

Jena

Chemnitz

Dresden

Liberec

Ústí nad Labem

10

110 miles (175km)

Hof

Chomutov

Karlovy Vary

Kolešov

55 miles (90km)

Hradec Králové

11

Cheb

06

02

PRAGUE

Ostrava

Mariánské Lázně

03

01

Beroun

05

09

Olomouc

Plzeň

Humpolec

Fürth

Brno

Regensburg

Jindřichův Hradec

07

Telč

SLOVAKIA

Deggendorf

04

Třeboň

08

Ingolstadt

12

Passau

Augsburg

Linz

St Pölten

VIENNA

BRATISLAVA

Munich

AUSTRIA

Most visitors to the Czech Republic only see touristy Prague and that's like ordering a pizza and only eating the pineapple. There's much to see beyond the metro lines: forested hills, forgotten historical towns and treks to ancient castles.

PRAGUE

● ARTS & CULTURE   ● HISTORY   ● OUTDOORS   ● FOOD & DRINK   ● FESTIVALS & EVENTS   ● MUSIC & FILM

——— ONE HOUR FROM ———

01 Plzeň

The industrial town of Plzeň (Pilsen in Germany) may not be high on most people's itinerary, but this is the home of Pils beer and the brewery is a must-visit. The informative tour is followed by an inevitable tasting session, after which – if you are still on your feet – you can peruse one of the best beer-themed museums in the world. *www.prazdrojvisit.cz; 1hr by train or bus.*

02 Křivoklát Castle

There are several well-known Czech castles that share the tourist limelight, but you will earn serious castle-bagging kudos from in-the-know locals when you say you're off to visit Křivoklát. It's arguably the most beautiful castle of them all – a Gothic hilltop fortress picturesquely lost among the wooded hills of Central Bohemia. This drowsy village and its towering bastion teleport you to a different world. *www.hrad-krivoklat.cz; 1hr 30min by train.*

———TWO HOURS FROM———

03 Mariánské Lázně

The Czech Republic has some wonderful spa towns, but whereas most people head to over-gentrified Karlovy Vary (Carlsbad), the smaller, friendlier Mariánské Lázně has become a hit with more discerning spa connoisseurs. On the verge of a Unesco listing and featuring a long line of illustrious guests, the exquisite 19th-century colonnades here are backed by a wall of forested hills, creating one of the most pleasant spa environments in the country. *2hr by train.*

04 Šumava National Park

Extending along the southwest border with Bavaria and Austria, the Czech Republic's wildest national park is an attractive prospect for anyone longing to leave the 21st century behind. The low forested mountains, sparsely populated and packed with wildlife, are a real outdoor playground, riddled with hiking and cycling trails. Come in June or September to avoid the peak summer months, though not many foreigners make it here. *www.npsumava.cz; 2hr by car.*

05 Litomyšl Castle

The town of Litomyšl is famous for its Unesco-listed Renaissance château, and also for being the birthplace of Czech composer, Bedřich Smetana. He was actually born in the château's brewery and would no doubt have visited the baroque palace theatre in the west wing, which is one of the best in the whole country. Away from the château, the baroque houses and uninterrupted arcades on Litomyšl's main square are a delight to explore. *www.zamek-litomysl.cz; 2hr by train.*

06 Slavkovský Forest

This huge, thickly forested area between Mariánské Lázně and Karlovy Vary is filled with low mountains latticed with hiking and cross-country skiing trails. Very sparsely inhabited, the region teems with wildlife and even harbours a couple of endemic plant species. The unofficial capital, Bečov nad Teplou, is a marvellously atmospheric place, its castle perched high on a rocky outcrop above the River Teplá. *2hr by train or car.*

RAILS TO THE TRAILS

Not only does the Czech Republic have one of the most dense rail networks on earth, it can also boast a world record when it comes to hiking trails. In no other country will you find so many marked routes, all of them meticulously maintained by volunteers from the Czech Hiking Club (www.kct.cz), which has branches in almost every town and publishes very detailed hiking maps. So, if you're looking to escape the tourist crush of Prague, take the train to the hills where all sorts of hiking routes await.

07 Telč

A short trip along the D1 motorway, Unesco-listed Telč is one of those Czech towns that have you wondering why 99% of tourists go to Prague. Known for its Renaissance architecture and château, the main square, with its arcading and ornate gables, is almost completely ringed by lakes and moats and has the feel of an open-air museum. *2hr by car.*

08 Třeboň

Hungry? Leave Prague's tourist restaurants behind for some real Bohemian fare in Třeboň, a town in South Bohemia known for its carp and locally brewed beer. The fish in question come from the murky waters of the countless artificial ponds and lakes outside the town, which supply the main ingredient of a Czech Christmas dinner – roast carp. The small Regent brewery serves up the south's finest lager. *2hr 30min by train.*

09 Olomouc

The eastern city of Olomouc is one of those places that's constantly touted as the next big thing – fortunately it's stayed below the radar. Moravia's answer to Prague has a wealth of baroque masterpieces, cobbled squares and a laid-back, studenty atmosphere. The must-sees here are the Unesco-listed Trinity column – allegedly the biggest single baroque sculpture in central Europe – and the fountains that grace the city's quaint piazzas. *2hr 30min by train.*

10 Sněžka

Shared between Poland and the Czech Republic, Mt Sněžka is the Czechs'

highest peak, but you don't need to be Tenzing Norgay to reach the top. Just over a couple of hours in the car brings you to Pec pod Sněžkou, from where marked trails lead 6.6km to the summit. Higher than the treetops at 1603m above sea level, the widescreen views are awe-inspiring. *2hr 30min by car.*

© Angelafoto / Getty Images

09

——— THREE HOURS FROM ———

 Colours of Ostrava Festival

The Czech music festival not to miss is Colours of Ostrava, held in mid-July in the former industrial city. In previous years this huge event has attracted acts as diverse as Björk, NERD and Grace Jones, as well as countless lesser-known artists from central and Eastern Europe. ***www.colours.cz; Jul; 3hr 10min by express train.***

12 Lipno Reservoir

In the country's deep south, this long, snaking reservoir on the River Vltava has become a very popular destination for locals who – far from the sea – come here to windsurf, paddleboard and generally mess about on the water. In summer you can camp on the reservoir's banks, and in winter hire cross-country skis and hit the trails through the dense forests. ***3hr by car.***

© Matyáš Theuer

LUXURIOUS BOLT

Dare to dream? Looking to splurge on something fancy for a special occasion? Here's where to live your best life in Europe.

HOLES IN EUROPE

Le Refuge, France

Shiny new Le Refuge is the highest mountain resort in France, with an altitude of 2551m. You're going to feel on top of the world every day, with 360-degree views. *www.lerefuge-valdis-ere.com/en; Sommet de Solaise, Val D'Isère; 2hr from Chambéry/ Grenoble airports by car.*

Six Senses, Portugal

It may be part of the luxurious Six Senses chain, but this 19th-century manor house, perched high in the hills overlooking Portugal's acclaimed wine country, the Duoro Valley, has a unique vibe. Relax in the Wine Library with advice from sommeliers who are winemakers themselves. *www.six-senses.com/resorts/douro-valley/destina-tion; Quinta de Vale Abraão, Samodães, Lamego; 1hr 20min from Porto airport by car.*

Stamba Hotel, Georgia

Bet you didn't have Tbilisi, Georgia on your luxury destination list. Time to rethink that! This grand escape even has its own chocolaterie and roastery for all your artisanal chocolate and coffee needs. Its plush casino is something straight out of a Wes Anderson film. *www.roomshotels.com/stamba-hotel; 14 Merab Kostava St, Tbilisi; 10min from Tbilisi centre by car.*

The Alpina, Switzerland

It's all about high-end Alpine chic in Gstaad. But one resort rises above the others. The Alpina's two-storey Panorama Suite with its own hot tub overlooking the mountains is frequently named among the world's best. *www.thealpinagstaad.ch; Alpinastrasse 23, Gstaad; 1hr 30min from Bern airport by car.*

Perivolas, Santorini, Greece

You want that Greek Islands fantasy life? Just to rub it in, they call them 'lifestyle houses' at Perivolas. These cliff-carved suites on Santorini's northern tip have endless ocean views and the best infinity pool you'll ever see. Yes to this life. *www.perivolas.gr; Oia, Santorini; 5hr from Athens by high-speed ferry.*

Sheen Falls Lodge, Ireland

Arriving at this remote former 17th-century fishing lodge overlooking Sheen Falls, where deer freely roam over 300 acres, is a little like stepping onto an otherworldly film set. Oenophiles will love the biggest privately owned cellar in Ireland, and the Wine Academy events. *www.sheenfallslodge.ie; Kenmare, Co. Kerry; 1hr 30min from Cork by car.*

Ca's Xorc, Mallorca, Spain

Mallorca's kept this treasure well hidden along a bumpy precipice, but the tranquillity is worth the tricky journey. A restored 18th-century olive mill set between the mountains and the sea, it brims with character as well as luxurious touches. *www.casxorc.com/en/cas-xorc-home; near Sóller, Mallorca; 40min from Palma airport by car.*

Il Sereno, Italy

With a location that redefines serenity, this remarkable hotel right on Lake Como's eastern shore would make even local resident George Clooney envious. There's a Michelin-starred restaurant and, if you fancy it, you can be picked up in a Maserati Quattroporte. *www.serenohotels.com/property/il-sereno; Via Torrazza 10, Torno; 50min from Milan airport by car.*

Hotel Adriatic, Croatia

You can't miss this beacon of loveliness in Rovinj's main square. Splurge on The Adriatic Suite with its two balconies on two different wings for views of St Katarina island and the lingering orange, pink and purple sunsets Croatia is famous for. *www.maistra.com/hotel-adriatic-rovinj; Obala Pina Budicina, Riva Pino Budicin 16, Rovinj; 40min from Pula airport by car.*

Inverlochy Castle Hotel, Scotland

It'll feel like you're in *Downton Abbey* at this 19th-century Scottish castle, and with the acclaimed Albert and Michel Roux Jr devising the cuisine, and furniture provided by the King of Norway, you might start pinching yourself that it's all a dream. *www.inverlochycastlehotel.com; Torlundy, Fort William; 1hr 30min from Inverness by car.*

Bjarkalundur
Varmahlíð
Þristapar
Efri-Brú
Hof
Laugarbakki
Laugar
Reykjaskóli
Stykkishólmur
Búðardalur
08
Skjöldur
ICELAND
Sandur
07
Hraunsnef
Hellnar
Ytri-Tunga
105 miles (170km)
70 miles (115km)
Borgarnes
35 miles (55km)
02
Saurbær
05
Melahverfi
06
Gullfoss
Akranes
Þingvellir
Geysir
Hrauneyjar
01
Laugarvatn
REYKJAVÍK
03
Sólheimar
Sandgerði
12
Hafnir
Núpar
Selfoss
Grindavík
Þorlákshöfn
Hella
04
Hvolsvöllur
11
Skógar
10
Vestmannaeyjar
09
Vík

NORTH ATLANTIC OCEAN

Iceland hosts an incredible number of visitors, yet many limit their adventures to the charming capital and/or bathing in the Blue Lagoon. Big mistake. This is a magical country of geographical contrasts and some epic natural phenomena.

REYKJAVÍK

● ARTS & CULTURE   ● HISTORY   ● OUTDOORS   ● FOOD & DRINK   ● FESTIVALS & EVENTS   ● MUSIC & FILM

———— ONE HOUR FROM ————

01 **Þingvellir National Park, Golden Circle**

Þingvellir is a big draw for history, geology and archaeology fans, and especially anyone familiar with Iceland's famous Sagas. This sprawling national park sits in a rift valley, formed by the separation of two tectonic plates, and houses Alþing, the site of Iceland's (and the world's first democratic) parliament in 930AD. The frozen-in-time aura is unmistakeable as you stroll around the stone foundations of the ancient encampments. *www.thingvellir.is/en; 45min by car.*

02 **Hótel Glymur**

For private hot tubs and spectacular views over Hvalfjordur (the 'whale fjord'), this bright-red hotel has split-level rooms ideal for watching Icelandic nature do its theatrical stuff. In summer, you can fish for salmon or trout in the nearby stream, take a two-hour hike to Glymur Waterfall (Iceland's second-highest at 198m) or go horse riding, before dining on langoustines in the candlelit restaurant. *www.hotelglymur.is; 50min by car.*

03 **ION Luxury Adventure Hotel**

This luxurious eco-retreat with its spaceship-like decor on stilts may raise some eyebrows, but those floor-to ceiling windows certainly maximise Northern Lights viewing, and staff will phone you in your room when the lights are visible. They can also arrange fly-fishing excursions, horseback riding and glacier tours, or you can just relax in the bar sipping a wild rhubarb liqueur, watching the mist roll over the lava fields. *https://ionadventure. ioniceland.is/the-hotel; 1hr by 4WD only.*

04 **Hotel Rangá**

South Iceland's only four-star resort revolves around enjoying nature, with popular dog-sledding and snow-mobile tours, as well as stargazing from the astronomical telescopes in its observatory. The well-travelled will get a kick out of the 44 continent-themed rooms: boomerangs and kookaburra stained-glass windows in the 'Australia' suite, or the 'Africa' suite fooling you into thinking a safari's on the cards outside, rather than the Northern Lights. *www.hotelranga.is; 1hr 20min by car.*

05 **Gullfoss, Golden Circle**

Fingers crossed for a sunny day, because that's when this gigantic waterfall is adorned with rainbows. Gulfoss, which translates as 'Golden Falls', is part of Iceland's 'Golden Circle' tour circuit and there are two stages to it: the straightforward path to the top of the falls or the windy trail down to the rugged canyon where the water gushes 32m down. *www.gullfoss.is; 1hr 30min by car.*

06 **Geysir Geothermal Area, Golden Circle**

Yes, that's steam coming out of the ground. It's 100°C underneath your feet. You'll soon see a crowd of people looking skywards with their cameras in the air, waiting for something to happen. Go join them to see 'Strokkur' in action: a geyser that explodes up to 20m into the air every 10 minutes or so. It's equally exciting and terrifying. *1hr 30min by car.*

RESPONSIBLE TOURISM

You'll see a lot of signage around Iceland advising you to be a responsible visitor. The sheer popularity of the country as a tourist destination, its small population and the fragile landscape requires respect for its sustainability. If you're not following a guided tour, always stick to official roads rather than damage the local flora or scar the delicate environment. Hiring a 4WD can be mandatory for some areas but that doesn't mean it comes with a licence to go off-road (in fact, it's illegal to do so). Always check weather forecasts and road closures before you set off.

—— TWO HOURS FROM ——

 07 **Snæfellsnes**

This remote, rugged peninsula has a name that sounds like a wicked *Lord of the Rings* character ('Sneye-fess-ness'). The otherworldly Snæfellsjökull Glacier, which inspired Jules Verne's *Journey to the Centre of the Earth*, is a photographer's paradise, as is the 19th-century wooden Búðir Church. Hike the grand landscapes then unwind at the heavenly Hotel Búðir, for one of the best gourmet Icelandic meals of your visit. *www.budir.is; 2hr by car.*

08 **Narfeyrarstofia**

Seafood is what it's all about in Stykkishólmur, a well-preserved fishing village in western Iceland at the northern tip of Snæfellsnes, and Narfeyrarstofia is one of the best places to eat it. The fresh daily catch comes from the adjacent Breiðarfjörður – you could be savouring local mussels in broth with home-baked bread, or seafood soup with crab, scallops and fish. Local lamb is also a highlight. *www.narfeyrarstofa.is; 11.30am-10pm; 2hr by car.*

09 **Slippurinn**

Destination dining klaxon! This family-run restaurant on the Westmann Islands is only accessible by ferry but definitely worth the trip (the same team spawned Reykjavík's Michelin Bib Gourmand-winning Matur og Drykkur). Slippurinn's focus on slow cooking and local, wild produce is very popular with fans of New Nordic cuisine. Book well ahead. *www.slippurinn.com; summer only; 2hr 30min by car (including 13km ferry journey).*

10 **Vik**

Visit Iceland's most southerly village to experience walking along the black-sand beach, Reynisfjara, at low tide. Admire the majestic basalt cliffs, teeming with birdlife in summer, including adorable puffins (just don't get too near the overly protective arctic terns in nesting season). Photographers will love dramatic Dyrhólaey nearby (especially at sunrise/sunset), a 120m-high headland that lives up to its name, meaning 'the hill island with the door hole'. *2hr 30min by car.*

Thórsmörk

Gather your powers for a hike in the 'Valley of Thor' in the Southern Iceland highlands. This secluded spot is under glaciers Eyjafjallajökull, Mýrdalsjökull and Tindfjallajökull, with appropriately awe-inspiring scenery, including contrasting lush greenery and lava fields. It's also challenging to reach, via gushing rivers. There's a range of rustic accommodation available to book in the Volcano Huts. *www.volcanohuts.com; super-jeep or guided tour recommended, May–Oct only; 2hr 30min by car.*

——— THREE HOURS FROM ———

Landmannalaugar

Translating as the 'people's pools', Landmannalaugar is another hiking hotspot. Its name is a clue to pack your swimming costume and towel for some geothermal hot-spring bathing. Bubbling mud pots, lava fields, otherworldly rhyolite mountains in shades of red, pink, green, blue and yellow? Tick, tick, and tick. *www.fi.is; guided tours with the Icelandic Touring Association mid-Jun to mid-Sep; 3hr by 4WD.*

HOT TUB TIPS

Bathing in Iceland's geothermal hot tubs is a once-in-a-lifetime joy but there is a certain etiquette involved before you commit to it. You must do as the locals do and shower thoroughly without a swimsuit beforehand. There are shower facilities for this purpose beside most outdoor hot springs and hot tubs. Iceland's natural pools have no chemicals, so it's important to keep them pristine. Icelanders are also used to being naked in front of each other, so don't be daunted to discover open showering facilities in same-sex changing rooms at public pools. No one looks, so nude up.

© Pyty / Shutterstock

N
0
0
150 km
90 miles

ITALY
135 miles (215km)

Livorno

Poggibonsi
Arezzo
Città di Castello
Ancona

Siena
Civitanova Marche

13
Perugia
18
90 miles (145km)
Assisi
16
Foligno
Ascoli Piceno

Orvieto
11
Teramo

Viterbo
45 miles (70km)
Rieti
Pescara

L'Aquila

Civitavecchia
04
12
Sulmona
05
Cerveteri
02
Tivoli
Avezzano
09
★
ROME
Frosinone
Campobasso

Anzio
Latina
14
Caserta
Terracina
15
10
Naples
17
Pompeii
Salerno

0
0
20 km
10 miles
★ ROME

Frascati
do di Ostia
Grottaferrata
01
08
06
03
Castel Gandolfo
07

ROME

Ancient and modern routes radiate from Italy's capital, taking you to hilltop castle towns, ancient acropolises, picturesque lakes, storied ocean views and plenty of unique wine and eating destinations.

⬤ ARTS & CULTURE ⬤ HISTORY ⬤ OUTDOORS ⬤ FOOD & DRINK ⬤ FESTIVALS & EVENTS ⬤ MUSIC & FILM

———— ONE HOUR FROM ————

01 **Castelli Romani**

Just 20km south of Rome, this easy getaway in the Alban Hills is well loved for its fine wines, easy access and excellent foods. The key to visiting the area is settling into the rhythms of this land that was once the holiday destination for popes and princes. As you move from the towns of Frascati, Castel Gandolfo and Lago Albano, don't miss a slow-cooked porchetta in Ariccia, the dry white wines of Frascati, and delicate wild strawberry tarts in the village of Nemi. **30min by car.**

02 **Tivoli**

Hilltop Tivoli can claim two Unesco World Heritage Sites. The Villa Adriana was used by Emperor Hadrian as a retreat in the 2nd century. The pools and immaculate gardens invite hours of exploration. Take a guided tour to really dive deeper into the history, art and culture of the Villa. Next door is the Villa d'Este. Built in the 16th century, this garden includes 50 fountains, 64 waterfalls and 220 basins – all run with ingenious gravity-fed systems. **30min by train or car.**

03 **Ostia**

On weekends, Romans flock to the coast for long lunches in the little seaside ristorantes and trattorias of the coastal village of Lido di Ostia. The seafood here is to die for, as are the endless snacks and unparalleled wine pairings in sophisticated *enoteche* (wine bars). Grab a seat and a Campari in one of the many beach clubs for endless views of the Mediterranean. **40min by car.**

04 **Lago di Bracciano**

As the Roman summer heats up, people turn their sights north for quick getaways to the crystal shores of Lake Bracciano. The volcanic lake offers up great hikes, sailing, canoeing, and some gravel-sand beaches – perfect for sunning yourself in between dips to cool off in the azure waters. **40min by car.**

05 **Santa Marinella**

While beach time rarely makes it on to most Rome itineraries, it's a wonderful diversion from the hustle and bustle of the city. Romans have been basking on the glittering beach of Santa Marinella for thousands of years. Most people just pick a good *stabilimenti* (beach club) and hang out on the beach of the arching bay for a blissful day. **40min by train.**

06 **Castel Gandolfo**

Overlooking the deep crater of Lago Albano, architectural explorers love wandering the streets of Castel Gandolfo. This tiny Lazio village is home to just 8900 residents. Architectural highlights include the Apostolic Palace of Castel Gondolfo, which houses a fine museum, as well as the church of St Thomas of Villanova. **40min by train.**

07 **Infiorata di Genzano**

Travellers in the know come here in June for the Corpus Domini celebration known as the Infiorata di Genzano, where locals pave the streets in ornate flower carpets depicting religious or historical scenes. The finale is a glorious parade followed by one of Italy's biggest flower petal fights. **Jun; 40min by car.**

WINE TOURS FROM ROME

Every village of Italy will have a unique assortment of wines you've probably never tried. When in doubt, just get the house wine – it's almost always fantastic. Travel north from Rome, passing the hilltop villages of the lower parts of Tuscany on a pilgrimage to sample what many people say is one of Italy's most famous wines: Brunello de Montalcino. On your trip through the Castelli Romani, be sure to sample the well-known whites from Frascati. Lazio is primarily known for its Trebbiano wines, including the celebrated whites from Orvieto.

© Henk Goossens / 500px., © trabantos / Shutterstock; © Alexandra Bruzzese

08 Ostia Antica
A wonderful day trip from Rome takes you to the well-preserved ruins at Ostia Antica. This archaeological site dates back to the 4th century BC, when it was founded as a fortified military camp to guard the mouth of the river Tiber. At its zenith, the city had a population of 50,000. *40min by train or car.*

09 Cerveteri
One of the country's most famous necropolises is located within easy driving distance of Rome. The 900-acre Necropoli della Banditaccia in the abandoned Etruscan town of Cerveteri houses over 1000 tombs, dating back to the 9th century BC. *50min by car.*

10 Naples
You haven't had pizza until you've had a carefully wood-fired slice in one of Naples' famed pizzerias – of course this is where pizza as we know it today was invented. Don't miss the chance to try a pizza alla Napoletana with anchovies, olives, pickles,

© StevanZZ / Getty images

© rarrarorro / Shutterstock

capers and oregano at Da Michele, which makes a cameo in the film *Eat, Pray, Love*. As you wander through Naples' colourful streets, visiting little wine shops and open-air markets, keep your eyes open for delicate pastries known as *sfogliatelle*. **www.damichele.net; *1hr 20min by train.***

11 Orvieto
Located in the neighbouring region of Umbria, Orvieto is another Italian city providing the double whammy of spectacular sightseeing and amazing food.

On a walking tour through town, leave at least an hour to explore the 14th-century duomo in the centre of the city. While the Gothic facade is extremely impressive, it's the detail and artistry of the interior frescoes that will leave a lasting impression, including Luca Signorelli's evocative depiction of the Last Judgement. End your day with a hike up to the top of the Torre del Moro, for spectacular views of the city. **1hr 30min by train or car.**

—— TWO HOURS FROM ——

12 **Lago della Duchessa**
Lago della Duchessa ('Lake of the Duchess') is the place to come for easy hikes at 1788m above sea level. That's high enough to remain pretty cool in summer, when Rome swelters. In winter, don snowshoes for a magical hike through a winter wonderland. **www.riservaduchessa.it; 1hr 40min by car.**

TRAVELLING TASTE BUDS

The best way to order in Italy is to find a local to order for you. Each village will have a farm-fresh ingredient, pasta dish or preparation technique they are famous for. In and around Rome, you'll definitely want to sample pecorino cheese, authentic Roman pasta sauces like carbonara and amatriciana (a spicy tomato and bacon sauce), distinctive thin-crust pizzas and classic *cucina romana* dishes featuring lamb, veal and offal. In the south, contrast the doughier Neapolitan pizza. Umbria is best known for its salamis, while further north in Tuscany, don't miss bistecca alla Fiorentina, delicate truffle sauces and perfectly paired red wines.

© Xinhua / Alamy Stock Photo

© Matt Munro

13 Umbria Jazz Festival

 Since 1973, the Umbria Jazz Festival has attracted some of the biggest names in the business to play the July festival in the Etruscan town of Perugia and surrounding cities in the Umbria region. Perugia is a tinderbox for culture, music and arts, with several universities located here, and the town's rich history evident in every cobbled alley, grand palazzo and buzzing piazza. *www.umbriajazz.com; Jul; 2hr by train or car.*

14 Reggia di Caserta

 Unesco calls this 18th-century palace, the largest royal residence in the world, 'the swansong of the spectacular art of the baroque, from which it adopted all the features needed to create the illusions of multidirectional space.' Expect marble, frescoes, monumental paintings and period furnishings inside; exquisite gardens outside. Movie fans will recognise several parts of the palace, which have featured in two *Star Wars* films. *2hr by car.* *www.lonelyplanet.com/italy/caserta*

15 Sperlonga

Known mostly as a beach resort, Sperlonga contains a surprisingly chic historic centre, with plenty of boutiques, galleries, cafes and restaurants to explore. In the 1950s this was a favourite retreat for Hollywood's glitterati. *2hr by car from Central Rome.*

16 Quintana Jousts

 Taking place in June and August, the jousts of La Quintana will take you back to the medieval heyday of this Umbrian city. You'll love the pageantry, costumes, parades and street festivals that precede the main jousting events. Alas, there's no knight-on-knight action here. Rather, the jousters try to pull a ring from a statute dedicated to the Roman God Mars. Impressive nonetheless. ***www. quintanadiascoli.it; Jun & Aug; 2hr by train or car from Central Rome.***

Pompei

The ruins of Pompei have an effect that transcends time. Perhaps it's the fact that the city wasn't just abandoned, but smothered in volcanic pumice from the erupting Mt Vesuvius in AD 79. This blanket of fire and dust preserved the city, and its streets, brothels and temples, to provide modern-day explorers with a glimpse into the world of the Roman Empire. ***2hr by train.***

18 Assisi

Italy is for pilgrims. Catholics and non-believers alike, however, will be floored by the power, beauty and spiritual energy of the birthplace of St Francis. While primarily known as a religious destination, this city – with its towering basilica, pristine medieval centre and views of plains below and mountain above – will entrance you at every turn. ***2hr 10min by car or 2hr 20min by train.***

BY ROMAN ROAD, BUS, BIKE, TRAIN

You can reach villages near and far by train from Rome's central Termini station. It's quick, easy, cheap, environmentally friendly, and these days the trains mostly come on time. If you know a bit of Italian, consider renting a car. Italy is also famous for its walking and cycling. From Rome, consider biking out to Appia Antica and Lago de Bracciano (www.bikemap. net). While you'll want to stick to side roads, many destinations close to Rome can easily be reached on two wheels or even on foot.

© Greg Elms

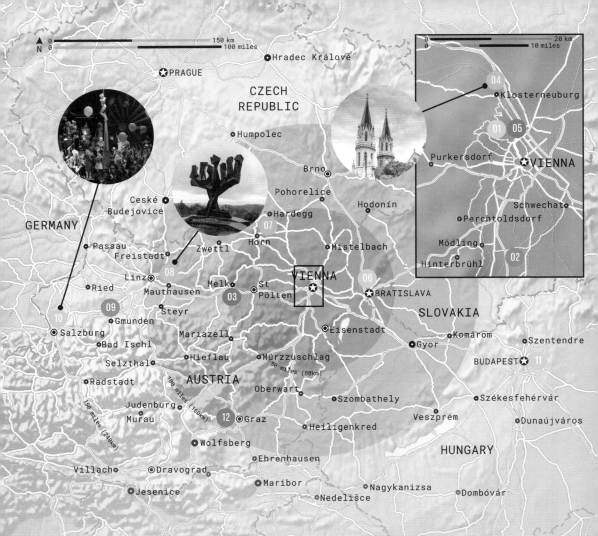

N
0 — 150 km
0 — 100 miles

0 — 20 km
0 — 10 miles

HRADEC KRÁLOVÉ

⭐ PRAGUE

CZECH REPUBLIC

Humpolec

Brno

Pohorelice

Hodonín

Hardegg

07

Ceské Budejovice

GERMANY

Passau
Freistadt

Zwettl

Horn

Mistelbach

VIENNA

⭐ VIENNA

06

🏙 BRATISLAVA

SLOVAKIA

Linz

08

Melk

St Pölten

Ried

Mauthausen

03

Eisenstadt

Gyor

Komárom

Szentendre

10

Steyr

Mariazell

Gmunden

09

⭘ Salzburg

Bad Ischl

Hieflau

Mürzzuschlag

Selzthal

50 miles (80km)

AUSTRIA

Radstadt

100 miles (160km)

Oberwart

Szombathely

Veszprém

Székesfehérvár

BUDAPEST ⭐ **11**

Dunaújváros

Judenburg

Murau

150 miles (240km)

12 ⭘ Graz

Heiligenkred

HUNGARY

Wolfsberg

Villach

⭘ Dravograd

Ehrenhausen

Maribor

Jesenice

Nedelišce

Nagykanizsa

Dombóvár

04

Klosterneuburg

01

05

Purkersdorf

⭐ VIENNA

Schwechat

Perchtoldsdorf

Mödling

02

Hinterbrühl

Escaping from Vienna's grand boulevards and monuments into the Austrian countryside is easy. You'll reach dense forest after just 20 minutes on the subway. Further afield, explore baroque treasures and sights in Hungary and Slovakia.

VIENNA

ARTS & CULTURE ⬤   HISTORY ◯   OUTDOORS ⬤   FOOD & DRINK ◯   FESTIVALS & EVENTS ◯   MUSIC & FILM ◯

———— ONE HOUR FROM ————

01 Grinzing

A typically charming 19th-century village just outside Vienna proper, Grinzing rewards visitors with gingerbread-house architecture. But what draws the crowds are the Heurigen: boisterous rustic taverns serving young wine, usually in a courtyard setting. The menus seem unchanged in centuries: roast pork, blood sausage, bread with lard, pickled vegetables and potato salad, and strudel for dessert. ***10min by tram from Wien Oberdöbling station.***

02 Laxenburg

Castle fans can enjoy one-stop sightseeing at Laxenburg, a vast estate that was used as a Habsburg retreat in the 14th century. On wooded grounds dotted with lakes and castles, you'll find a network of walking paths, some several kilometres long. The gardens near the entrance gate get crowded in summer, but a 15-minute walk further on yields serenity. ***www.schloss-laxenburg.at; 30min by bus from Südtiroler Platz-Hauptbahnhof.***

03 Melk Abbey

Perched on a rocky outcrop, the blockbuster abbey-fortress Stift Melk is a high point of any visit to the Danube Valley. The monastery church, with its twin spires and high octagonal dome, dominates the 500-room complex. The interior is baroque gone barmy, with regiments of smirking cherubs, gilt twirls and polished faux marble. Ride a scenic river ferry to Krems and return to Vienna from there. ***www.stiftmelk.at; 1hr by train.***

04 Klosterneuburg Monastery

Another of Austria's enormous baroque monasteries, Klosterneuburg overlooks the Danube from its medieval foundations. Don't miss the Verduner Altar, which dates to the 12th century and has reliefs moulded in copper plating. For liquid pursuits after visiting the monastery, the surrounding town has excellent wineries and cafes, some serving wines from the monastery's own vineyards. The hills nearby offer head-clearing hikes amid fruit trees. ***www.stift-klosterneuburg.at; 1hr by train & tram.***

05 Beethoven Museum, Heiligenstadt

One of dozens of places where Beethoven laid his head in and around Vienna, this apartment endures today as a museum. It's set in the once-isolated spa town of Heiligenstadt on the city's northern edge. Beethoven wrote his lament about going deaf here in 1802, as well as works including *Piano Sonata No 2* and the first versions of the *Third Symphony*. ***www.wienmuseum.at/en/locations/beethoven-museum; 1hr by train.***

06 Bratislava, Slovak Republic

Bratislava's castle presides over a pastel-hued old town, which is crowned by Michael's Gate and Tower: climb past the five small storeys of medieval weaponry in Bratislava's only remaining gate (hailing from the 13th century) for superior old-town views. Then stroll along St Michael's St and sample a range of restaurants and cafes, offering hearty food and wide selections of beer at very low prices. ***1hr 10min by train.***

THE LOCAL'S VIEW

'Vienna has a plethora of parks both big and small where you can leave the city chaos behind for a while, but none top the Wienerwald, the Vienna Woods. They are easily reached by public transport, so you can be in thick woods and far away from the noise and hurry of Vienna before you know it – a perfect place to rejuvenate the soul and breathe in the sweet air from Vienna's green lungs. Afterwards, look for a traditional Heurigen (wine tavern), where you'll find a mix of *gemütlich* (comfortable) atmosphere and carafes of wine.'

Neal Bedford, diplomat

© Johannes Simon / Getty Images; © manfredxy / Shutterstock; © volkerpreusser / Alamy Stock Photo

──── TWO HOURS FROM ────

 Hardegg Castle
Unlike so many 'castles' that are really palaces – with huge windows and fine gardens but a distinct lack of defences to guard against an invading horde – Burg Hardegg looks altogether more fortress-like. Perched on a knoll overlooking the River Thaya, the castle's foundations date back to the 12th century. On tours, you can clamber through the fortifications. *www.burghardegg.at; 2hr by train.*

08 **Mauthausen Concentration Camp**
At Mauthausen concentration camp, prisoners were forced into slave labour in a granite quarry. More than 100,000 died or were executed here between 1938 and 1945. Now an affecting memorial, visitors can walk through the remaining living quarters and see the gas chambers. A display area shows charts, artefacts and photos of prisoners and their SS guards. *www.mauthausen-memorial.org; 2hr 20min by train.*

09 **Gmunden**
With its lakeside setting, pretty Gmunden exudes a breezy, Riviera feel. Once known primarily for ceramics and the salt trade, today it's a fashionable and very scenic retreat. Cafes and restaurants line the waterfront, main square and Marktplatz. The real joy here is hiking along the dozens of routes that follow the Traunsee shore and lead into the hills of the Traunstein. *www.gmunden.at; 2hr 20min by train.*

10 **Salzburg Festival**
The highlight of the Salzburg events calendar is the summertime Salzburg Festival. It's a grand affair that features about 200 productions – including theatre, classical music and opera – staged in the huge 1950s Grosses Festspielhaus, the historic Haus für Mozart and the baroque Felsenreitschule. Tickets vary widely in price and you'll need to book them far in advance as this is a major international draw. *www.salzburgerfestspiele.at; Jul & Aug; 2hr 30min by train.*

© Getty Images

—— THREE HOURS FROM ——

Great Market Hall,
Budapest, Hungary

Budapest's Great Market Hall (Nagycsarnok), completed in 1897, more than lives up to its name. Locals and tourists come here for fruit, vegetables, deli items, fish, dried meats and sausages, plus as many kinds of honey as you'd care to name. Head up to the first floor for Hungarian crafts, including painted eggs and embroidered tablecloths.
www.piaconline.hu; 2hr 40min by train.

12 Austrian Open Air Museum, Stübing

Approximately 100 Austrian farmstead buildings fleck the verdant grounds of the Austrian Open Air Museum (Österreichisches Freilichtmuseum) in Stübing, near Graz. It's ideal for a family outing as there are all manner of treats for kids, including cute animals to pet and muddy streams for splashing about in. The stout structures have been gathered from across Austria and are fun to explore.
www.freilichtmuseum.at; 3hr by train.

© Alex Segre / Alamy Stock Photo

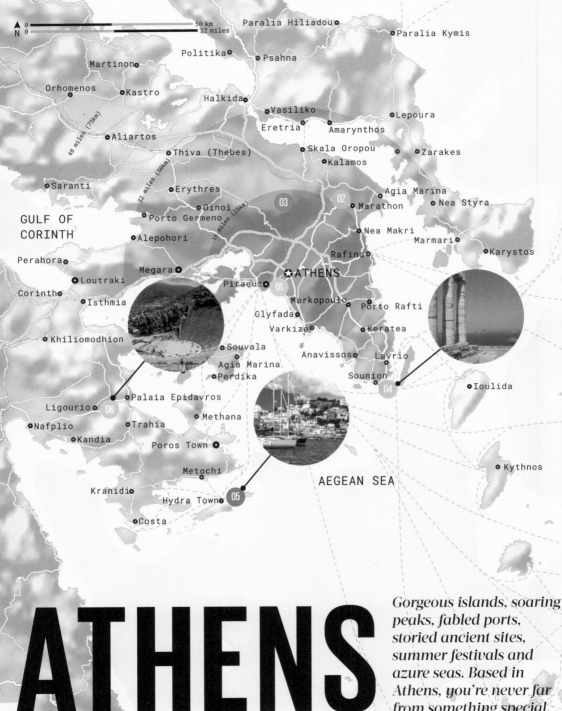

Paralia Hiliadou

Paralia Kymis

N
0 ____ 50 km
0 ____ 32 miles

Politika

Martinon

Psahna

Orhomenos

Kastro

Halkida

Vasiliko

Lepoura

48 miles (75km)

Aliartos

Eretria

Amarynthos

Zarakes

Thiva (Thebes)

Skala Oropou

Kalamos

Saranti

32 miles (50km)

Erythres

Agia Marina

Nea Styra

Oinoi

03

02

Marathon

GULF OF
CORINTH

Porto Germeno

16 miles (25km)

Nea Makri

Marmari

Alepohori

Rafina

Karystos

Perahora

Megara

Piraeus

✪ ATHENS

Loutraki

01

Corinth

Isthmia

Markopoulo

Porto Rafti

Glyfada

Khiliomodhion

Varkiza

Keratea

Souvala

Anavissos

Lavrio

Agia Marina

Sounion

Perdika

04

Palaia Epidavros

Ioulida

Ligourio

06

Methana

Nafplio

Trahia

Kandia

Poros Town

05

Metochi

Kythnos

Kranidi

Hydra Town

AEGEAN SEA

Costa

ATHENS

Gorgeous islands, soaring peaks, fabled ports, storied ancient sites, summer festivals and azure seas. Based in Athens, you're never far from something special.

● ARTS & CULTURE   ● HISTORY   ● OUTDOORS   ● FOOD & DRINK   ● FESTIVALS & EVENTS   ● MUSIC & FILM

ONE HOUR FROM

01 **Seafood dining, Piraeus**
Piraeus is the biggest port in the Mediterranean and the hub of the Aegean ferry network. The Great Harbour is backed by lots of gritty cafes and fast-food joints, but with a little effort you'll find gems serving the local seafood bounty, such as calamari, shrimp and mullet. Look around Mikrolimano, Zea Marina and along the waterfront promenade at Freatida. The port restaurant Margaro is superb. *30min by metro.*

02 **Marathon Battlefield**
The small town of Marathon was the site of one of the most celebrated battles in world history in 490 BC. Road signs mark the historic route of Pheidippides, the courier who ran to Athens to announce victory, and thus gave the name to the 42km race. The battlefield itself, 4km south of the town, has a tomb for the 192 men killed, a model of the battle and historical information. *8am-3pm Tue-Sat; 1hr by car.*

03 **Mt Parnitha National Park**
Mt Parnitha comprises many peaks, the highest of which is Karavola (1413m), tall enough to get snow in winter. Hiking trails weave through forests and arid landscapes pocked with caves, and the park is popular for mountain biking as well. The easiest way to explore is on the path through Tatoi, the 40-sq-km grounds of the former summer palace. There is much wildlife in the park, including the European red deer, golden eagles, hawks and owls. *www.parnitha-np. gr; 1hr 10min by car.*

TWO HOURS FROM

04 **Temple of Poseidon, Cape Sounion**
Talk about location. At Cape Sounion, the Temple of Poseidon stands on a craggy spur that plunges 65m to the sea. Built in 444 BC – the same year as the Parthenon – it's a vision of gleaming white columns. Sailors in ancient times knew they were nearly home when they saw their first glimpse of white. *odysseus.culture.gr; 2hr by bus from the KTEL Mavromateon Terminal.*

05 **Hydra**
Enchanting Hydra is the one Greek island free of wheeled vehicles. There are just tiny marble-cobbled lanes, donkeys, rocks and sea. The island's exquisitely preserved stone architecture, criss-crossing rural paths and clear, deep waters have attracted artists throughout the ages. Look for high-season art shows at the Melina Mercouri Exhibition Hall and in small galleries across the old town. *http://ydra. gov.gr; 2hr from Athens by ferry from Piraeus.*

06 **Hellenic Festival, Epidaurus**
Built of limestone, yet one of the best-preserved Classical Greek structures in existence, this grand 4th-century BC theatre is the undisputed highlight of Epidaurus. Renowned for its amazing acoustics – a coin dropped in the centre can be heard from the highest seat – the theatre seats up to 14,000 people and is used for performances of Ancient Greek drama during the annual Hellenic Festival. *www.greekfestival.gr; Jul & Aug; 2hr by car from Athens.*

THE LOCAL'S VIEW

'A favourite getaway is Mt Parnitha for a bird's-eye view of the whole Athenian basin. Densely forested, it's the ideal place to hike and enjoy the local rakomelo (honey-infused liquor) at the shelter, especially during winter. I suggest using the cable car for a scenic ride to the top. It leaves you inside the Regency Casino Mont Parnes and you then walk stairs and corridors leading to mountain trails. Or, try the small town of Arachova on top of Mt Parnassus, about 170km from Athens. It's the perfect destination for winter sport aficionados.'

Dimitris Foussekis, artist

© nicolesy / Getty Images; ©rfave / Budget Travel; ©Freeartist / Getty Images

EUROPE'S BEST BACK

No phone signal? No problem. Recalibrate your overworked mind and body in sea, fo

Cycling Through Water, Belgium

The name is exactly what it is, but you don't get wet. What?! Imagine a 212m-long, 3m-wide concrete bike trail through the pond of the De Wijers nature reserve in Bokrijk-Genk. You're at eye level with the water and the swans. *www.cyclingthrough-water.com; Genk; 1hr from Brussels by car.*

Treehotel, Sweden

Try to contain your inner-child's excitement when you see the artistically designed treehouses suspended 4–6m off the ground in this remote Swedish pine forest. You could choose to sleep in the UFO treehouse, the Bird's nest or the Mirrorcube. *www.treehotel.se/en; Edeforsväg 2A, Harads; 1hr 10min from Stockholm by plane.*

Linden Tree Retreat & Ranch, Croatia

Want to sleep in a Native American tipi in the Croatian wilderness? A Scandinavian-style mountain cabin? An African mud hut with a cast-iron stove? Or just a safari tent on the riverbank? Here, within Velebit Nature Park, you can. *www.lindenretreat. com; Velika Plana; 2hr from Zagreb/Split by car.*

Chewton Glen, England

Forget real life in one of the seven secluded luxury treehouses at Chewton Glen, at the edge of Hampshire's New Forest. Suspended 10m off the ground, soak in your outdoor hot tub on the deck and have food hampers delivered via a hatch. *www.chewtonglen. com/stay-over/tree-houses; New Milton, Hampshire; 2hr 30min from London by car.*

Manhausen Island, Norway

Location, location. This remote part of northern Norway is 55 acres of serenity in the middle of the the Grøtøya strait. Relax in the saltwater dam overlooking the Lofoten mountains, or in the architecturally designed cabins with glass windows jutting into the sea. *www.manshausen.no/ en; Lille Manshausen; 1hr 30min from Bodø Airport to Nordskott by ferry, then boat transfer.*

TO-NATURE RETREATS

sky and mountains. Here are Europe's best places to unwind in nature.

Brenners Park Hotel & Spa, Germany

Perfectly located for walks through the Black Forest, this luxurious resort in an historic park and botanic garden stretches along the banks of the river Oos in the centre of Baden-Baden.
www.grandluxury-hotels.com/hotel/brenners-park; Schillerstraße 4/6, Baden-Baden; 1hr 20min from Stuttgart by car.

La Ferme du Soleil, France

One for the snow bunnies, this chalet is only accessible by skis during snow season, and is the *dernier mot* in mountain tranquillity. You're so high up here, you'll be looking down on the clouds. In summer, it's all wildflowers, cows and mountain biking.
www.lafermedusoleil.com; Le Grand-Bornand; 1hr 20min from Geneva airport by car.

Eyvindara, Iceland

How about watching the Northern Lights from a hot pot in remote Eastern Iceland? You'll be within hiking distance of Fardaga Waterfall, as well as Lagarfljót, which is renowned for its mythical lake monster.
www.eyvindara.is; Egilsstadir; 3hr 20min from Akureyri by car.

Eleonas, Crete, Greece

At the foot of Mt Psioliritis, Crete's highest mountain, you'll find the village of Zaros, home to the island's spring-water supply and this gorgeous taverna with cottages set among the olive groves. It's the perfect base for hiking and mountain biking.
www.eleonas.gr; Zaros, Crete; 1hr from Iraklio by car.

Perdue Hotel, Turkey

Tricky to find, joyous when you do. You're on the edge of Babadağ Mountain here, surrounded by pine forests, and with the Mediterranean right outside your room (aka a deluxe canvas tent on wooden stilts).
www.perdue.com.tr/old/en; Uzunyurt Village, Kizilcakaya, Faralya; 1hr from Fethiye by car.

N
0 _____ 100 km
0 _____ 70 miles

Quillan
FRANCE
Vicdessos
Vielha
Perpignan
Benasque
Prades
Castejón de Sos
Boí
Le Boulou
Port-Vendres
18 17
Aínsa
Sort
Andorra la Vella
Puigcerdà
15
Adrall
La Molina
14
Cadaqués
Barbastro
Organyà
Bagà
Figueres
16
Benabarre
Tremp
Berga
Ripoll
Besalú
09
Ponts
Solsona
Gironella
05
11
Binefar
Cardona
Girona
08
10
Alfarràs
Vic
12
Balaguer
07
Tàrrega
Manresa
Vidreres
04
Lleida
Igualada
Tossa de Mar
Blanes
Torrente de Cinca
Terrassa
Granollers
02
SPAIN
Montblanc
Barcelona
Valls
01 Sitges
03
Tarragona
06
Miami Platja
13

Sant Carles de la Ràpita
MEDITERRANEAN
SEA
Vinaròs

Torreblanca
Ciutadella de Menorca

Benicàssim (Benicasim)
Maó

In the enchanting areas around Barcelona you'll find more of what you get in the city: world-class art, beautiful beaches, and gourmet food and wine to linger over. So you can follow Dalí's footsteps, feast on fresh seafood and work on your tan.

BARCELONA

● ARTS & CULTURE ● HISTORY ● OUTDOORS ● FOOD & DRINK ● FESTIVALS & EVENTS ● MUSIC AND FILM

—— ONE HOUR FROM ——

01 Sitges

When the weather's warm, catch a train down the coast to the classic resort of Sitges. Perhaps best known as a gay-friendly holiday destination with a rollicking Carnaval and a buzzing nightlife scene all summer long, Sitges also has a charming (and easily walkable) historic centre, an interesting fine dining scene and, of course, wonderful beaches fringed with tall palm trees. *50min by car or train.*

02 Montserrat Monastery

You can't just walk up to the door of this historic monastery – you have to take a cable car to the mountaintop on which it's located. Inside, you'll find a 12th-century sculpture of the Black Madonna, patron saint of Catalonia. If you're lucky, you might overhear the in-house boys' choir practising. The monastery museum has works by Dalí, El Greco and Monet and extending from this sacred spot is the Parc Natural de la Muntanya de Montserrat. *www.abadiamontserrat.net; 1hr by car or 2hr by train.*

03 Tarragona

The port city of Tarragona dates all the way back to the 5th century BC. The Roman ruins of Tarraco, including an amphitheatre, citadel and necropolis, were declared a Unesco World Heritage Site, but the nearby Cyclopean walls are believed to be even more ancient. Another highlight is Tarragona's Romanesque Catholic cathedral; there are nice beaches nearby, too. *www.catedraldetarragona.com; 1hr 20min by car or 1hr 40min by bus.*

04 Castell de Tossa de Mar

Castell de Tossa de Mar (also known as Vila Vella) is one of the most interesting of the Girona region's historic castles: once a fort built to defend against pirates, it's the only remaining example of a fortified medieval village along Catalonia's coast. Dating from the 12th century, the castle features a series of towers, some cylindrical and topped with machicolation openings where attackers could be fought off from atop the walls. There are spectacular sea views around the castle. *www.infotossa.com; 1hr 20min by car or train from Liceu.*

05 Banys Àrabs

Absorb some aquatic history away from the beach at the Banys Àrabs. The design of these historic baths was based on Islamic bathhouses mixed in with elements of Roman tradition. The complex features an elegant octagonal pool, a frigidarium (with cold water), a tepidarium (with warm water) and a sauna-like caldarium. You can tour all of these, plus the old apodyterium (changing room). *www.banysarabs.org; 1hr 20min by car or train from Liceu.*

06 Camí de Ronda

What could top a pleasant afternoon stroll along the Camí de Ronda, near the beach town of Salou? The 5.6km-long oceanfront path, originally built to allow fishers to observe the water below, offers gorgeous views over crashing waves and the rocky coastline. The walkway takes you to the quiet Cap de Salou, a great spot for birdwatching and relaxing. *1hr 30min by car.*

PLANES, TRAINS & AUTOMOBILES

If you're planning to drive, it's easiest to pick up a rental car at the airport when you land in Barcelona. If you'd rather get around the region by train or bus, you'll be travelling with RENFE (www. renfe.com) for longer-distance trains, and TMB (www.tmb.cat) for local trains, as well as city buses and the metro. If you're not in a hurry, or if budget is an issue, consider taking a bus instead of a train to destinations outside Barcelona. Check www. busbud.com, where all the bus companies' schedules are listed.

© Yurina_Photo / Shutterstock

 Cap de Sant Sebastià

The magical cape of Sant Sebastià features an historic 19th-century lighthouse, a medieval watchtower and an Iberian archaeological site in the old village of Sant Sebastià de la Guarda. Make the most of this particular stretch of coast by following the coastal path between Llafranc and Calella or Tamariu. Spend the night, if you have time, in delightful Calella de Palafrugell. *1hr 30min by car or 3hr 10min by train.*

08 **Gala Dalí Castle**

Salvador Dalí bought this medieval property in 1968 with the intention that his wife, Gala, would have a private retreat. Indeed, she spent summers there for most of the 1970s – and he only visited her when granted permission. After Gala's death in 1982, she was buried on-site, and Dalí himself moved in for a period. Today, it's open to the public as a museum. *www.salvador-dali.org/en/museums/gala-dali-castle-in-pubol; 1hr 30min by car.*

09 **Besalú**

Besalú is a well-preserved medieval town with an historic Jewish neighbourhood to explore, outdoor cafes and restaurants to enjoy, and surrounding mountains to hike. Just 90 minutes by bus from Barcelona, it's a perfect low-key escape, demanding little more than strolling around and sipping sangria on old stone terraces – ideal for a leisurely day trip, an overnight stay, or a whole weekend. *1hr 30min by bus or car.*

———— TWO HOURS FROM ————

10 **Cala d'Aiguafreda**

This rocky cove along the Camí de Ronda coastal path has crystal-clear water and beautiful views, not to mention a couple of simple seafood restaurants where you might just have the best meal of your trip. Try the breezy outdoor tables at Hostal Sa Rascassa (also a hotel) where the catch of the day tops the daily menu, and the mussels are to die for. *www.hostalsarascassa.com; 1hr 40min by car.*

11 Dalí Theatre & Museum

Can't get enough Dalí? The surrealist painter (1904–1989) left an indelible mark on the region, but it was here in Figueres where it all began and ended – he even designed the museum that holds his tomb. They're both located on the grounds of the theatre where Dalí's early works were first exhibited. *www.salvador-dali.org/en/museums/dali-theatre-museum-in-figueres; 1hr 40min by car or 2hr 20min by train.*

12 Aiguablava

Considered one of the finest beaches in Catalonia, Aiguablava (also known as Aigua Blava) sits on a bay near Begur and Palafrugell. With turquoise waters and tall cliffs, it's postcard-pretty, and a great place to sail, swim, or feast on freshly caught seafood. The nearby area is fairly well-developed, with a diving centre, rental equipment for water sports, public bathrooms and showers, shops and restaurants. *1hr 40min by car.*

SPRING
FLINGS

Visiting Barcelona in April? You're in luck: not only are Easter celebrations particularly lively in the region, but they're followed up by a week-long flamenco festival at Parc del Fòrum. Technically, it's in the city, but this waterfront complex isn't a typical tourist attraction. The colourful Feria de Abril (April Festival) has more than dance and music – expect a street fair-style festival with food vendors, horse-riding displays and enough paella to feed a crowd.

© Shutterstock / Simona Bottone

13 Platja del Fangar

Get off the beaten track at Platja del Fangar, a pristine beach on the Ebre Delta that stretches for more than 8km. Walk or take a bike ride along the shore, watch industrious fishers at work, go for a swim near the lighthouse, or just find a quiet spot to soak up some sun: you might not see another tourist. Bring food and a beach umbrella. *2hr by car.*

14 Catalonian Pyrenees

If you're visiting Barcelona in winter, a quick trip to the ski slopes of the Catalonian Pyrenees is hard to resist. With so many skiing opportunities elsewhere in Europe, this region is relatively relaxed (read: no waiting at the lifts). La Molina, the area's first ski resort, is a good option for families and beginners, while La Masella has more advanced options and excellent night skiing. *www.lamolina.cat, www.masella.com; 2hr by car.*

15 Casa-Museu Salvador Dalí

Between 1930 and 1982, Salvador Dalí lived and worked on and off at this quirky seaside residence with maze-like hallways, unusually shaped rooms and big picture windows facing Portlligat Bay. Book ahead for a tour through the house, which has since been turned into a museum; afterwards, stop into the small village of Portlligat (aka Port Lligat), located on the Cap de Creus peninsula. *www.salvador-dali.org; 2hr 10min by car or 3hr 10min by bus.*

16 Compartir

It's not hard to find a good meal in Barcelona, but there's something extra

special about a destination restaurant like Compartir, which means 'to share' in Spanish. It's a couple of hours' drive up the coast in Cadaqués, but you'll be rewarded with an open-air terrace table where the cava flows freely and sophisticated seafood is prepared by several chefs who trained at El Bulli. *www.compartircadaques.com; 2hr 10min by car.*

———— THREE HOURS FROM ————

17 Caldes de Boí

With 37 different natural springs, the spa of Caldes de Boí earned a spot in the Guinness World Records for its wide variety of thermal waters. Sign up for a spa treatment, soak in the healing baths or hole up at one of the grand hotels. In winter, take

advantage of the opportunity to ski or snowboard at Boí Taüll, the highest-altitude resort in the Pyrenees. *www.caldesdeboi.com; 3hr 20min by car.*

18 Vall de Boí

The Unesco World Heritage Site of Vall de Boí consists of eight Romanesque churches and a chapel. Spread out across the valley and surrounded by mountains, all were built during the 11th and 12th centuries following northern Italian designs. It's well worth a trip for architecture fans, especially if you have enough time for a leisurely stroll through the ancient villages where the churches are located. Pick up a map or join a guided tour at the interpretation centre, Centre del Romànic de la Vall de Boí. *www.centreromanic.com; 3hr 20min by car.*

© Fernando Fernández Baliña / Getty Images

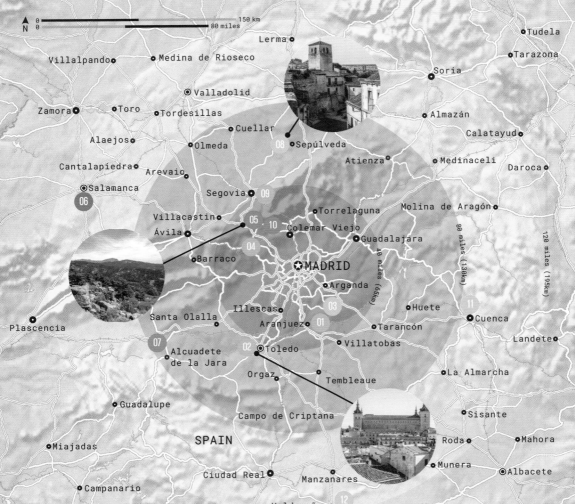

MADRID

Smack in the middle of Spain, Madrid is surrounded by historic towns, culinary hotspots and vineyards, made closer by good roads and rail links. In hot weather, it seems the whole city escapes to the hills, where crisp air and hiking await.

Map labels

Tudela
Tarazona
Soria
Lerma
Villalpando
Medina de Rioseco
Valladolid
Almazán
Zamora
Toro
Tordesillas
Calatayud
Cuellar
Alaejos
Olmeda
Sepúlveda
Atienza
Medinaceli
Daroca
Cantalapíedra
Arevaio
Salamanca
Segovia
Torrelaguna
Molina de Aragón
Villacastin
Colemar Viejo
Ávila
Guadalajara
Barraco
MADRID
Arganda
Huete
Cuenca
Santa Olalla
Illescas
Tarancón
Landete
Plascencia
Aranjuez
Villatobas
La Almarcha
Alcuadete de la Jara
Toledo
Orgaz
Tembleaue
Guadalupe
Campo de Criptana
Sisante
Roda
Mahora
Miajadas
Munera
Albacete
Campanario
Ciudad Real
Manzanares
Valdepeñas

SPAIN

150 km
80 miles

40 miles (65km)
80 miles (130km)
120 miles (195km)

● ARTS & CULTURE   ● HISTORY   ● OUTDOORS   ● FOOD & DRINK   ● FESTIVALS & EVENTS   ● MUSIC & FILM

──── ONE HOUR FROM ────

01 **Palacio Real, Aranjuez**

This opulent palace was created as a royal retreat, away from the hustle and bustle of Madrid; consider it an 18th-century great getaway for kings. See it on a guided tour, which highlights the palace's history and art collection. Even better is a stroll around the extensive gardens (entrance is free), which showcase native and exotic species. *www.patrimonionacional.es; 50min by train from Atocha, 50min by bus from Estación Sur or 50min by car.*

02 **Toledo**

Amid a tangle of narrow streets overlooking the Río Tajo, Toledo's old town has a wealth of churches, synagogues, mosques and museums to explore. They provide a glimpse back to Toledo's heyday, when it was a flourishing symbol of multicultural medieval society. The imposing cathedral, a symbol of the city's historical significance as the heart of the Catholic Church in Spain, is not to be missed. *www.catedralprimada.es; 50min by train from Atocha or 1hr by car.*

03 **Chinchón**

A charmingly ramshackle place, the village of Chinchón feels a world away from its powerhouse big city neighbour. Soak up the old-time atmosphere with lunch at one of the *mesones* (taverns) on the Plaza Mayor, which is lined with sagging balconies. Try Café de la Iberia, which has seating in an internal patio, but come early if you want to score a table on the balcony overlooking the square. *www.cafedelaiberia.com; 1hr by car.*

04 **San Lorenzo de El Escorial**

Rising from the foothills of the mountains that shelter Madrid from the north and west, the San Lorenzo de Escorial palace and monastery complex makes for a refreshing escape from the city heat. Completed in 1584, the complex became a centre of learning and the arts; inside, the walls are lined with masterpieces of art and tapestries and the barrel-vaulted library is one of the monastery highlights. *www.patrimonionacional.es; 1hr by bus from Intercambiador de Moncloa.*

05 **Cercedilla**

Easily accessible via the Cercanías train line from Madrid, the hiking trails in the foothills near Cercedilla make for the perfect city escape. Grab a map from the tourist office and follow one of a series of well-marked trails through fragrant pine forests, filling your lungs with the pure, crisp air. The *miradores* (viewpoints) trail offers stunning, panoramic vistas over the surrounding countryside. *www.cercedilla.es; 1hr 10min by train from Atocha.*

06 **Museo de Art Nouveau & Deco Casa Lis, Salamanca**

Nestled in the Salamanca city walls is the modernist mansion Casa Lis, a rare example of industrial architecture used in a residential building. Behind the striking facade of iron and glass lies a worthwhile collection of sculpture, paintings and art deco pieces, including Lalique glass and Limoges porcelain. This beautiful city has a buzzing nightlife, so stay overnight if you can. *www.museocasalis.org; 1hr 30min by train from Chamartín or 2hr 10min by car.*

AREAS OF NATURAL BEAUTY

It's easy to leave the cacophony of Madrid behind and enter the wilderness via several off-the-beaten-track nature spots within easy reach of the capital. Hiking trails traverse the Parque Natural Cañón del Río Lobos, a rocky river canyon near Burgos (2hr 10min by car). North of Segovia is the Parque Natural de las Hoces del Río Duratón (1hr 40min), a protected area surrounding a gorge that's a prime spot for birdwatching; the high rocky outcrops are nesting places for griffon vultures and Egyptian vultures; golden eagles and peregrine falcons have also been glimpsed nearby.

07 Vía Verde de la Jara

Spain's Vía Verdes (greenways) are a series of disused railway lines that have been transformed into cycling and walking trails. Covering 52km from Calera y Chozas to Santa Quiteria, the Vía Verde de la Jara is the closest to Madrid and one of the most picturesque (but not the best maintained; be sure to bring lights for the dark tunnels). The landscape is particularly beautiful in autumn. ***www.viasverdes.com; 1hr 30min by car.***

08 Sepúlveda

With houses staggered along a ridge carved out of a gorge, the picturesque town of Sepúlveda is a favourite weekend escape for Madrileños, who come to feast on the town's speciality dish of roast lamb. One of the best places to try it – roasted in a wood-fired oven – is Figón Zute el Mayor, which has been owned by the same family since 1850. ***www.figondetinin.com; 1hr 30min by car.***

09 Segovia

The great Roman aqueduct slicing through the heart of Segovia is an astonishing feat of engineering. Not a drop of mortar was used to hold together the uneven granite blocks that form the awe-inspiring arches, which rise 28m above the Plaza de Azoguejo. The city's Gothic cathedral and fairy-tale alcázar (castle) are equally worthy of your time. Meat-lovers shouldn't miss the local speciality, *conchinillo asado* (roast suckling pig). ***1hr 30min by car or 1hr 50min by train from Chamartín.***

10 Navacerrada

In winter, the Sierra de Guadarrama hills to the north of Madrid transform into a snowy paradise for skiers. The ski station at Navacerrada, which has lifts, and skiing and snowboarding gear rental, is within comfortable commuting distance of the capital. There's access to a number of slopes, ranging from easy to difficult, and non-skiers can take a walk beneath snowy pines. ***www.puertonavacerrada.com; 1hr 30min by car or 1hr 50min by train from Atocha.***

——— TWO HOURS FROM ———

11 **Holy Week, Cuenca**
Semana Santa (Holy Week) is celebrated throughout Spain, but the parade in Cuenca is one of the most evocative. The occasion is marked by an eerie, silent procession through town by local brotherhoods dressed in coloured robes and peaked hoods; the silence is interrupted only by the occasional drumbeat and bugle call. *2hr by car or 2hr 10min by bus.*

12 **La Bodega de las Estrellas, Valdepeñas**
Surrounded by vineyards, the otherwise unremarkable town of Valdepeñas is a destination for wine-lovers. One of several local wineries is La Bodega de las Estrellas ('the winery of the stars'), a family-run business that produces organic Tempranillo, Cabernet Sauvignon and Syrah. It offers wine tastings and courses in wine astrology. *www.labodegadelasestrellas.com; 2hr by car or 2hr 10min by bus.*

ON THE TRAIL OF CERVANTES

South of Madrid, Castilla-La Mancha is the land traversed by knight-errant Don Quixote in Miguel de Cervantes' 16th-century masterpiece. Several trails claim to feature places referenced in the work. Dulcinea, the platonic love of Quixote, is honoured at the Casa-Museo de Dulcinea in El Toboso (1hr 30min by car); another Cervantes sight is the Museo de Quijote y Biblioteca Cervantina in Ciudad Real (1hr by train from Atocha or 2hr 10min by car). Finally, in honour of the knight who spent his time tilting at windmills, the molinos de viento near Consuegra (1hr 20min by car) make for the perfect Don Quixote shot.

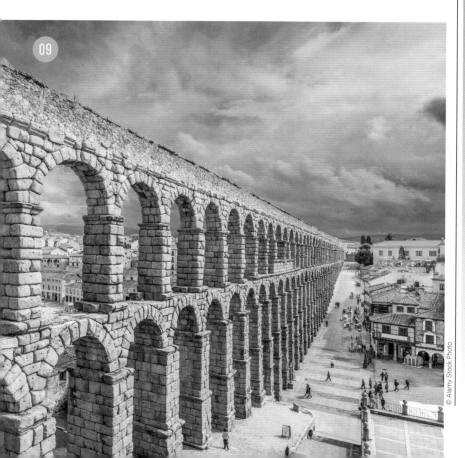
09

BALTIC SEA

◉ Kiel

14
17 ◉ Rostock
◉ Lübeck
◉ Schwerin
◉ Hamburg
08
10

16

18

◉ Koszalin

POLAND

15

05

◉ Szczecin

04

GERMANY

01 ☆ BERLIN
02 **03**

◉ Magdeburg

◉ Gorzów Wielkopolski

13
◉ Poznań

◉ Zielona Góra

06

◉ Cottbus

◉ Göttingen

07 ◉ Leipzig

◉ Legnica
◉ Wrocław

◉ Erfurt **12**
11 ◉ Jena

◉ Chemnitz

09 ◉ Dresden

◉ Liberec
◉ Ústí nad Labem

CZECH REPUBLIC

☆ PRAGUE

◉ Hradec Králové

N

0   200 km
0   100 miles

55 miles (90km)
110 miles (175km)
165 miles (265km)

BERLIN

Surprisingly close to the German capital are verdant parks and Baltic coast beaches. Across the region are examples of Germany's history, and culture high and low.

● ARTS & CULTURE ● HISTORY ● OUTDOORS ● FOOD & DRINK ● FESTIVALS & EVENTS ● MUSIC & FILM

——— ONE HOUR FROM ———

01 Grunewald

Grunewald, an upper-crust Berlin suburb, contains gardens, parks, palaces and a handful of museums. The Grunewald forest, a vast fresh-air refuge criss-crossed by trails and dotted with lakes extending all the way west to the Havel River, offers respite from the city. Wild boar, deer and other animals make their homes here. Nature paths fan out from the Grunewald S-Bahn stop. *20min by S-Bahn.*

02 Sanssouci Palace & Park, Potsdam

This glorious park and palace ensemble was dreamed up by Frederick the Great (1712–86) and is anchored by the eponymous palace – Frederick's favourite summer retreat, somewhere he could be 'sans souci' (without cares). It's an expansive place, lavishly beautiful inside and out. Take your time wandering along the park's meandering paths. *www.spsg. de; 30min by train.*

03 House of the Wannsee Conference, Wannsee

In January 1942, a group of 15 high-ranking Nazi officials met in a stately villa near disconcertingly bucolic Lake Wannsee to hammer out details of the 'Final Solution', the systematic deportation and murder of European Jews. The chilling exhibits displayed in the rooms where the conference took place document the meeting and the racial persecution leading up to it. *www.ghwk.de; 30min by suburban train, then bus 114.*

04 Choriner Musiksommer, Chorin Abbey

This romantically ruined monastery near a little lake and surrounded by a lush park was built by Cistercian monks in the 13th century. It is widely considered one of the finest red-brick Gothic structures in northern Germany. The open-air summertime series of classical concerts, Choriner Musiksommer, features top orchestras and soloists from around Germany, with up to 20 concerts per season. *www.choriner-musiksommer.de; Jun-Aug; 50min by train.*

05 Ravensbrück Concentration Camp, Fürstenberg

Some 90km north of Berlin, Ravensbrück was one of Nazi Germany's first concentration camps. It primarily housed women, especially Jewish women and those from Poland. Upwards of 130,000 people were imprisoned here during WWII; few survived. Today the site has some remaining buildings as well as memorials and museums. Displays recall the trains used to transport prisoners. *www.ravensbrueck.de/ mgr/neu; 1hr 10min by train.*

06 Lübbenau, Spreewald

Poet Theodor Fontane called Lübbenau the 'secret capital' of the Spreewald, the vast area of forests, canals, rivers and wetlands southeast of Berlin. The pretty little town can get deluged by day trippers, but simply stroll away from the main street to escape the crowds. Better yet, take a punt boat tour along the narrow channels past lovely countryside, old mills and inviting inns. *www.spreewald.de/en/ boat-rides; 1hr 10min by train.*

THE LOCAL'S VIEW

'Take a trip to Mecklenburg Lake Plateau – the 'land of a thousand lakes' – or Hamburg. Though you'd think it's just another big city, it's completely different from Berlin. Check out Hamburg's Elbe River area with its culture, cafes and even tiny beach bars. If the bunch of local lakes and a river don't feel watery enough for you, then visit the island of Rügen located in the real (Baltic) Sea.'

Oliver Naatz, illustrator

© MNTravel / Alamy Stock Photo; © ullstein bild / Getty images; © picture alliance / Getty images

07 Leipzig

Leipzig is Saxony's coolest city, a playground for young creatives who have been displaced from fast-gentrifying Berlin. It's also a city rich in history and culture. For centuries, it was the heart of German publishing and its annual book fair dates back to 1632. The fair attracts hundreds of authors, thousands of publishers and over 150,000 literary fans who link up at nearly two thousand events. *www.leipziger-buchmesse.com; late-Mar; 1hr 20min by train.*

08 Hamburg

Hamburg's passion for music hasn't missed a beat since The Beatles hit the city in the early 1960s, and remains deep in its core today. From orchestral overtures in the stunning new Elbphilharmonie concert hall, to intimate, sweaty clubs in St Pauli and beyond, where you'll hear great new bands, plus all sorts of venues in between, Hamburg is Germany's premier destination for live music. *1hr 20min by train.*

── TWO HOURS FROM ──

09 Dresden

There are few city silhouettes more striking than Dresden's. The spires, towers and domes belonging to palaces, churches and stately buildings only hint at the treasures within. Foremost among them is the New Green Vault (Neues Grünes Gewölbe) in the 15th-century Renaissance city palace (Residenzschloss), which displays over 1000 dazzling objects wrought from gold, ivory, silver, diamonds and jewels. *www.skd.museum; 2hr by train.*

© Thomas Quack / Shutterstock

10 Schwerin Palace & Gardens

Gothic and Renaissance turrets, Slavic onion domes, Ottoman features and terracotta Hanseatic step gables are among the mishmash of architectural styles that make up Schwerin's inimitable schloss, which is crowned by a gleaming golden dome. Outside, the surrounding Burggarten (castle garden) brings you to the baroque, manicured Schlossgarten, intersected by several canals. *www.mv-schloesser.de/de/location/schloss-schwerin; 2hr by train.*

11 Bratwurstmuseum, Amt Wachsenburg

Bratwurst, or sausage, is one of Germany's national foods. There are hundreds of local versions enjoyed across the country, from tiny finger-sized sausages grilled over an open flame in Nuremberg, to the large and substantial creations of Thuringia. This small museum celebrates all things bratwurst. Although it's said you never want to see sausages being made, here you'll get the

chance! *www.bratwurstmuseum.de; 2hr 30min by train.*

12 Weimar

The historical epicentre of Germany's 18th-century Enlightenment, Weimar is an essential stop for anyone with an interest in the country's history and culture. A pantheon of intellectual and creative giants lived and worked here: Goethe, Schiller, Bach, Cranach, Liszt, Nietzsche, Gropius, Herder, Kandinsky – and the list goes on. You'll see them memorialised on the streets, in museums and in reverently preserved houses across town. *2hr 30min by train.*

——— THREE HOURS FROM ———

13 Poznań

Poland's architectural gem, Poznań, is centred on one of central Europe's finest city squares, the Rynek. Dozens of Gothic, Renaissance and baroque buildings line the four sides, their facades painted in a palette of warm pastels that come together in one vast beautiful mural. In the middle stands the 16th-century town hall, with its striking 61m-high tower. More treasures await in the surrounding narrow lanes, including museums and attractive old fountains. *2hr 50min by train.*

HIGH-SPEED TRAVEL

Germany's comprehensive train network makes car-free travel from Berlin easy. High-speed lines optimised for the vaunted ICE fast trains radiate out towards Hamburg, west towards the Rhine and south to Nuremberg, bringing compelling destinations within a three-hour radius of the capital. Alternatively, the legendary autobahns, with their lack of speed limits, allow quick jaunts out of town. Also worth noting is that on slower, traditional train lines, there is both no need (and indeed no option) for reservations. Frequent service (usually hourly) means you can travel on a whim.

14 Warnemünde

One of Germany's most popular seaside resorts, Warnemünde still has the feel of the fishing village it once was, its streets lined with charming little cottages, many now housing restaurants. Go for long walks along the Baltic on the wide and startlingly white beaches and frolic in the gentle surf on summer days. Later, relax over a refreshing German draught beer at an open-air cafe or snack on some locally caught fish sold at stalls across town. *2hr 50min by train.*

15 Müritz National Park

Müritz National Park is an oasis of lakes and forests in the heart of otherwise uninterrupted farm country. There are well over 100 lakes here, as well as countless ponds, streams, bog and wetlands. Boardwalks and other features scattered throughout the park let you get close to nature – look out for ospreys, white-tailed eagles and cranes. The serene, light-dappled beech forests are recognised by Unesco. *www.mueritz-nationalpark.de; 3hr by car.*

16 Stralsund

Historic Stralsund's sea-scented air and Gothic architecture make a very appealing combination. Wander the cobbled streets, heading towards the lively port area. Smoked-fish stands dot the harbour; many have quite elaborate menus but nothing costs more than €5. The smooth, oily fillets of northern fish are smoked for hours until they have a tangy, buttery softness that melts in your mouth. *3hr by train.*

15

16

Bad Doberan Cathedral

The former ducal residence of Bad Doberan, near the Baltic coast, was once the site of a Cistercian monastery. Construction of this magnificent Gothic church started in the 13th century. Treasures inside include an intricate high altar and an ornate pulpit. A massive restoration project has made every one of the cathedral's 1.2 million bricks look like new. ***www.muenster-doberan.de; 3hr by train.***

 Usedom

Nicknamed 'Badewanne Berlins' (Berlin's Bathtub) in the pre-war period, Usedom Island is renowned for its 42km stretch of beautiful white-sand beach. Its eastern tip sits just across the border in Poland. Long hours of sunshine make this the sunniest place in Germany. Explore along one of the many woodland bike and hiking trails, and admire the elegant 1920s villas with wrought-iron balconies. ***3hr 20min by train.***

CELEBRATE SUMMER

Summer festivals are sacred to the German soul – although there's nothing particularly sacred about these fresh-air bacchanalias. Even the smallest of villages holds a summer festival, usually featuring litres of locally brewed beer and regional wines quaffed at long tables. Popular foods are served in great quantities – expect plenty of the local version of bratwurst. Entertainment ranges from traditional drinking songs and ballads performed by troupes in costumes to big-name acts, in the case of large cities, and small-time bands (often fresh from rehearsing in the garage) in little towns.

18

N

0 150 km
0 100 miles

SWITZERLAND

Delémont Zürich

Solothurn

Neuchâtel Lucerne
BERN Altdorf
Andermatt

Lausanne

Geneva Sion Mörel
Bellinzona

Chamonix

Aosta

Modane Ivrea Novara
Rivoli 03
Oulx Casale Monferrato Asti
Alessandria

FRANCE

Savigliano Alba

Cúneo

Saint-Auban Genova 04
San Remo Savona

Nice MONACO
Cannes

GERMANY

Bregenz

Innsbruck

VADUZ

AUSTRIA

150 miles (240km)

Zernez Merano Cortina
D'Ampezzo

Edolo Belluno

Trento Conegliano

100 miles (160km)

ITALY Vicenza
06 09
01 08 Venice
Bergamo Brescia Verona 11
Monza Erbusco
Milan 07
Pavia Cremona Poggio Rusco
02 Piacenza
Parma 05
Reggio Modena
Emilia
Pievepelago Forlì
Rapallo Sestri Levante SAN MARINO
La Spezia 12 Pistoia
Luccia 10
Pisa Firenze

Livorno Arezzo

Poggibonsi Siena

50 miles (80km)

MILAN

From Milan, getaways beckon. The Italian lakes, the Alps, historic coastal towns and the plains of northern Italy offer year--round adventures. Italy's good train network puts it all within easy reach.

● ARTS & CULTURE   ● HISTORY   ● OUTDOORS   ● FOOD & DRINK   ● FESTIVALS & EVENTS   ● MUSIC & FILM

——— ONE HOUR FROM ———

01   **Bergamo**

Bergamo is one of northern Italy's most beguiling cities. Head for the ancient hilltop Città Alta (Upper Town), a tangle of tiny streets, encircled by 5km of Venetian walls. It features a wealth of art and medieval, Renaissance and baroque architecture, a privileged position overlooking both the southern plains and the newer (and lower) town, and some fine dining. *1hr by train.*

02   **Certosa of Pavia**

One of the Italian Renaissance's most notable buildings is the splendid Certosa di Pavia. Giangaleazzo Visconti of Milan founded the monastery, 10km north of Pavia, in 1396 as a private chapel and mausoleum for the Visconti family. This underrated treasure is a unique hybrid between late-Gothic and Renaissance styles and is well worth a visit.
www.certosadipavia.com; 1hr by train.

03   **Palazzo Borromeo gardens, Lake Maggiore**

The busiest of the Lake Maggiore islands, Isola Bella has the vague appearance of a vessel, with the grand Palazzo Borromeo at the prow and the gardens at the rear. There are 10 tiers of spectacular terraced gardens here, a fine example of baroque Italian landscaping. White peacocks strut about this exquisite landscape, as can you.
www.isoleborromee.it/eng/isola-bella; 1hr by train.

04   **Pesto, Genoa**

It would be criminal to come to Genoa and not try pesto genovese. The city's famous pasta sauce – a pounded mix of basil, pine nuts, pecorino and parmesan, olive oil and garlic – really does taste, and look, better here than anywhere else. It's traditionally served with trofie, Liguria's distinctive, twirled pasta. Places to savour this treasure range from corner delis to fabled restaurants. *1hr 30min by train.*

05   **Prosciutto, Parma**

If reincarnation ever becomes an option, pray you come back as a citizen of Parma. Where else can you enjoy fresh-from-the-attic prosciutto made from pigs who spend their lives gorging on another local delicacy, parmigiano reggiano. On largely car-free streets lined with Romanesque and Renaissance gems, you'll find all manner of cafes, trattorias and small markets where you can savour silky, salty prosciutto.
1hr 30min by train.

06   **Lake Lugano**

Spilling over into Switzerland from northern Italy, Lago di Lugano is a sparkling blue expanse that can be enjoyed from the transport hub of Lugano. Less overrun than other lakes straddling the border, it is bewitching, whether glimpsed from one of the many trails that wind along its shores or from the deck of the ferries that glide across it. *1hr 30min by train.*

——— TWO HOURS FROM ———

07   **Opera Festival, Verona**

Around 14,000 music-lovers pack Verona's Roman arena on summer nights during the annual opera festival, which

MAKE TRACKS FOR NORTHERN ITALY

Much of the region around Milan, especially to the north, is covered not by Trenitalia, Italy's national railway, but by Trenord, an independent operator whose tracks serve hundreds of towns and cities across the region. It offers train passes that are valid from one to seven days for prices from €16.50 to €43.50. These are remarkable deals as they cover not just Trenord's hundreds of trains but every other public transport in the region, including trams, buses, cable cars and even many ferries. As an added bonus, the pass lets you ride first class on trains.

draws international stars. The setting is unforgettable. Built of pink-tinged marble in the 1st century AD, Verona's Roman amphitheatre has morphed into the city's legendary open-air opera house. Book ahead for crowd-pleasing classics like *La Traviata*, *Aida*, *Tosca* and *Carmen*. **www.arena.it; late Jun–late Aug; 1hr 40min by train**.

08 Franciacorta wine region

The Province of Brescia in Lombardy is renowned for Franciacorta, a naturally sparkling wine that comes in both red and white varieties. It's known for its long ageing process – longer than Champagne. This sunny region has a unique soil, rich in minerals. Among the wineries, one of the most respected is Ca' Del Bosco, which welcomes visitors year-round. **www.cadelbosco.com; 2hr by train.**

09 Villa Carlotta, Tremezzo, Lake Como

Visit lakeside Tremezzo to savour one of Lake Como's loveliest sights, the 17th-century Villa Carlotta. The art-packed interior is eclipsed by the sprawling botanic gardens outside. The 8 hectares are bursting with colour from orange trees interlaced with pergolas, together with the blooms of some of Europe's finest rhododendrons, azaleas and camellias. Access it all from one of the delightful lake ferries. **www.villacarlotta.it; 2hr by train.**

10 Duomo, Florence

Florence's Duomo is the city's most iconic landmark. Capped by Filippo Brunelleschi's famed red-tiled cupola, it's a staggering construction whose breathtaking pink, white and green marble facade and graceful campanile (bell tower) dominate the Renaissance cityscape. In the echoing interior, seek out frescoes by Vasari and Zuccari and 44 stained-glass windows. **www.ilgrandemuseodelduomo.it; 2hr by train.**

11 Carnevale, Venice

Masqueraders party in the Venice streets for the two weeks preceding Shrove Tuesday. A spectacular waterborne parade on the Cannaregio Canal gets the celebrations going. Tickets to nightly balls are costly, but there's no shortage of less expensive diversions, from costume competitions in Piazza San Marco to canal flotillas. All you have to do is wear a mask in order to be carnival-ready. **www.carnevale.venezia.it; Feb–Mar; 2hr 30min by train.**

CLASSIC CUISINE

Despite its relatively modest size, Northern Italy's cuisine has had a remarkable impact on our eating habits, filling restaurant menus and home-cooking repertoires. From the seafood of Venice to risotto alla Milanese, the pesto and focaccia of Genoa to the hearty meat sauces of Bologna, these regional Italian foods are now beloved the world over. Fast trains mean you can enjoy a classic lunch and unique wines, plus stock up on local foods and produce, on an easy – and extremely tasty – day trip.

—— THREE HOURS FROM ——

12 **Cinque Terre**
Set amid dramatic coastal scenery, these five ingeniously constructed fishing villages can bolster the most jaded of spirits. Sinuous paths traverse seemingly impregnable cliffsides, while a 19th-century railway line cuts through a series of coastal tunnels linking the villages (or 'five towns'), where cars are banned. A whole network of spectacular trails allow village-to-village hikes – come out of peak season to avoid hiking bottlenecks.
www.cinqueterre.eu.com; 3hr by train.

UNDERRATED ART M

Sure, you've ticked off The Louvre, the Van Gogh Museum, the Tate, etc, etc... but you honestly don't know what you're missing. Here are the quiet achievers.

USEUMS OF EUROPE

Laveronica Arte Contemporanea, Sicily

Sicily's not just about the lemons and volcanoes, you know. Check out this contemporary art gallery in Modica with a strong focus on politics and socially oriented art. The space is on the ground floor of an historic building and features Mediterranean artists. *www.gallerialaveronica.it; Via Grimaldi, Modica; 10am-1pm, 4-8pm Tue-Sat.*

Plan B, Romania

Located, appropriately enough, in Fabrica de Pensule ('the paintbrush factory'), this is both an exhibition space for contemporary art and a research centre unearthing Romanian art from the past 50 years. They also have a permanent exhibition space in Berlin. *www.plan-b.ro; Fabrica de Pensule, Str Henri Barbusse 59-61, Cluj-Napoca; 4-7pm Tue-Fri.*

Miva Fine Art Galleries, Sweden

Welcome to Sweden's largest international pop art gallery. It also has branches in Gothenburg and Stockholm but the headquarters here in Malmö is where you can discover signed limited edition graphics by Bob Dylan, for example. *www.mivagallery.se; Engelbrektsgatan 18, Malmö; hours vary.*

Villa Noailles, France

This modernist, palatial house-turned-gallery was built in Hyères, on the Mediterranean coast, in the 1920s for Charles de Noailles and his wife Marie-Laure, friends/supporters of Jean Cocteau, Pablo Picasso, Man Ray and Salvador Dalí, to name a few. It hosts fashion, photography, music, architecture and art events. *www.villanoailles-hyeres.com; Montée de Noailles, Hyères; 1-6pm Wed-Thu & Sat-Sun, 3-8pm Fri.*

Rarity Gallery, Mykonos, Greece

It's all about Greek hospitality from the moment you enter this early 20th-century Mykonian manor house in Hora (aka Mykonos Town). Accept an intimate tour of the five rooms showing contemporary international artwork while enjoying air-conditioned respite from the island heat. *www.raritygallery.com; Kalogera 20-22; 10am-midnight.*

Temnikova & Kasela, Estonia

Going strong since 2010, this Tallinn gallery wants to bring Eastern European art to a broader audience. Located in a Stalinist-era apartment, it has a stately grandeur and is actually Estonia's only art gallery; the founders are pretty much cultural celebrities. *www.temnikova.ee; Lastekodu 1, Tallinn; 3-7pm Wed-Sat.*

Museum of Urban & Contemporary Art, Germany

Europe isn't short of contemporary museums, but have you been to Munich's MUCA? Designed by street artist Stohead, the facade is an artwork itself, where you enter through a calligraphy 'door'. Its Mural restaurant mixes up fine dining with rough street-art ambience. *www.muca.eu; Hotterstrasse 12, Munich; 10am-8pm Wed-Mon.*

Lascaux International Centre for Cave Art, France

Where modern technology meets famous cave art: this full-scale replica of the entire Lascaux cave in the Vézère Valley was unveiled in 2016. It uses the best laser technology and 3D printing to allow visitors to experience what the (almost 600) original cave paintings were like. *www.lascaux.fr; Montignac; 9am-7.30pm (hours vary by season).*

Stavanger KunstMuseum, Norway

This lovely museum on Stavanger's Mosvatnet lake holds over 2600 works of Norwegian art, from the 18th century to the present day. Fans of landscape painter Lars Hertevig will find Norway's largest collection here. *www.stavangerkunstmuseum.no; Henrik Ibsens gate 55, Stavanger; 11am-4pm Tue-Sun.*

National Museum, Poland

With a choice of three permanent galleries here at MNK, the popular name for the main branch of Kraków's National Museum, you need to schedule your time wisely. Use the website to create an individual visit plan filtered by topic, artist or famous works. *www.mnk.pl; Al 3 Maja 1, Kraków; 10am-6pm Tue-Sat, to 4pm Sun.*

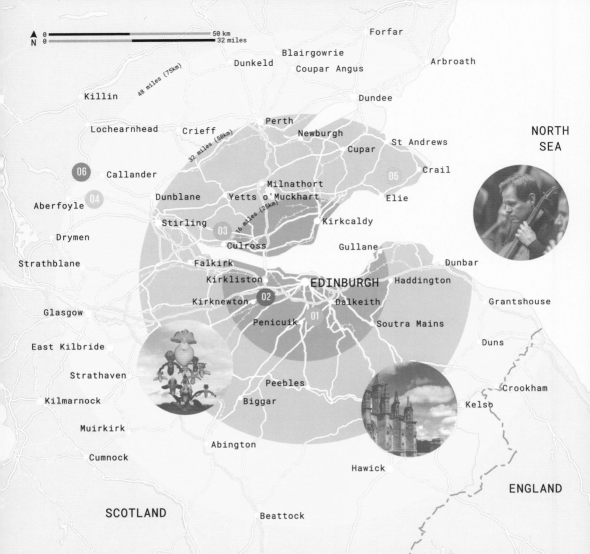

Forfar

Blairgowrie
Dunkeld   Coupar Angus   Arbroath

Killin   Dundee

Lochearnhead   Crieff   Perth   St Andrews
Aberfoyle   Newburgh
Callander   Cupar   Crail
Dunblane   Milnathort   Elie
Stirling   Yetts o'Muckhart
Culross   Kirkcaldy
Falkirk   Gullane   Dunbar
Kirkliston   Haddington   Grantshouse
Kirknewton   EDINBURGH
Penicuik   Dalkeith   Soutra Mains   Duns
Glasgow   Crookham
East Kilbride   Kelso
Strathaven   Peebles
Kilmarnock   Biggar
Muirkirk
Cumnock   Abington   Hawick   ENGLAND
SCOTLAND   Beattock

Aberfoyle
Drymen
Strathblane

NORTH
SEA

With its medieval lanes and Georgian new town, Edinburgh is no urban jungle. But travel a little further north and the scenery is epic. Historic towns, an outdoor gallery, innovative music and a foodie lakeside hotel are all within easy reach.

EDINBURGH

ARTS & CULTURE   HISTORY   OUTDOORS   FOOD & DRINK   FESTIVALS & EVENTS   MUSIC & FILM

——— ONE HOUR FROM ———

01 Rosslyn Chapel

Conspiracy theories swirl at this weird and wonderful 1446 chapel in the wee village of Roslin. Inside, every inch is decked out in ornate carvings. The most famous are the apprentice pillar, said to prefigure the helix structure of DNA, and a sculpted cob of corn, which predates Columbus' voyage to America where corn was 'discovered'. Go figure.
www.rosslynchapel.com; 20min by car or 50min by bus from Princes St.

02 Jupiter Artland

Semi-industrial hinterland is an unlikely setting for an artistic marvel. But enter the gates of this private outdoor art gallery for a magical world of sweeping sculpted earthworks, bronze weeping girls, a mystical void by Anish Kapoor and much more. This is a truly organic collection that is still evolving: the owners have commissioned 60 artists to date.
www.jupiterartland.org; 30min by car or 40min by bus from Princes St.

03 Culross

This snoozy village is crammed with grand 16th- and 17th-century buildings, reflecting its former status as a coal- and salt-mining centre and port. The tall town house was once a prison for presumed witches, and the ochre palace is painted and pine-panelled inside. Have a wander and lunch in the historic Red Lion – just remember to pronounce the place 'Koo-Ross'. *50min by car or 2hr by bus from Princes St.*

04 Lake of Menteith Hotel

This hotel enjoys the most tranquil of locations, looking out at the wind-ruffled waters of Scotland's only lake (all the others are lochs). Lunch on Scottish produce – super-local lake trout, Perthshire venison and hedgerow bramble crumble – then take a little ferry boat across the water to the romantic abbey ruins of one of the lake's islands. *www.lake-hotel.com; 1hr 10min by car or 50min by train to Stirling then 50min bus.*

05 East Neuk Music Festival

The quirkily monikered Neuk derives its name from the old Scots word for nook, and this cosy cluster of Fife fishing villages is now becoming known for a fantastic music festival, held over five days in July. Intriguing venues include a former nuclear bunker, caves and an aircraft shelter. The music itself runs from chamber to modern jazz and from avant-garde to classical.
www.eastneukfestival.com; Jul; 1hr 20min by car.

06 Ben Ledi

City streets seem a million miles away when you're tackling the slopes of one of Scotland's many mountains. You can take your pick in the gorgeous Trossachs, a lush region of lochs, peaks and genteel tourist towns. Ben Lomond provides the most famous hike, but the ridge of Ben Ledi is easier, and arguably more scenic. Ben Ledi *(overleaf)* is a conspicuous peak in the area, being visible from the Forth Bridge and even Edinburgh Castle.
1hr 30min by car or 50min by train to Stirling then 50min drive.

THE LOCAL'S VIEW

'Escape for me is a cobweb-blasting cycle ride or country walk with friends. The John Muir and West Highland Ways are a short drive away, meaning stages can be completed in a day or weekend. On long summer days I can leave work and in 20 minutes be on my bike and out of the city, heading to the nearby hills or along the East Lothian coast. Winter is all about skiing: the slopes of five Scottish ski centres are around two hours' drive away. If I'm short of time, I escape the city within the city, hiking the dramatic crag of Arthur's Seat.'

Clare Reid, research and innovation director

0 100 km
0 70 miles

11 70 miles (115km)
12 95 miles (150km)

N

08 Tampere
Kangasala

Pori

Vesilahti

Huittinen

06 Toijala

FINLAND

Sysmä
Pertunmaa
Onkiniemi
Puumala
Ristiina
Heinola
Vääksy
Tuulos
Lahti
Lappeenranta
Humppila
Forssa
Hämeenlinna
Taavetti
Kouvola
Mäntsälä
Virolahti
Aura
Koski
Somero
Hyvinkää
Forsby
Hamina
Karkkila
Kotka
09
Turku
Salo
02 Porvoo
07
Pargas (Parainen)
Perniö
Espoo
03
01
HELSINKI

GULF OF FINLAND

Nagu
(Nauvo)
Dragsfjärd
04
Karis (Karjaa)
Ekenäs (Tammisaari)

BALTIC SEA

Hanko **05**

Kunda

10 TALLINN
Jõhvi
Kiviõli
Paldiski
Keila-Joa

ESTONIA

Kärdla
Risti
Kauksi
Alajõe
Rohuküla
Märjamaa
Türi
Mustvee
Heltermaa
Kirbla
Jõgeva

105 miles (170km)

70 miles (115km)

35 miles (55km)

Finland is all about serene parks and walkable historic towns, plus giant hippo-like creatures espousing kindness and bravery. When Finns get out of town, they go to summer cottages on the lake, where they read, fish and relax in the sauna.

HELSINKI

● ARTS & CULTURE　　○ HISTORY　　● OUTDOORS　　◐ FOOD & DRINK　　◐ FESTIVALS & EVENTS　　● MUSIC & FILM

────── ONE HOUR FROM ──────

01 **Heureka Science Museum**
If you've got a) kids and b) time, before you fly out of Helsinki's Vantaa airport, stop by the Heureka Science Museum, where you can get in touch with your inner self. Literally. Its 'Wind in the Bowels' exhibit is exactly what it sounds like. *www.heureka.fi; 40min by car or bus.*

02 **Porvoo**
The postcard-worthy cuteness of Porvoo's historical town centre is practically tailor-made for strolling. The second-oldest city in Finland (established in 1380, after Turku around 1300) is famed for its picturesque 18th-century houses and its historic museum, Holm House, which displays all the accoutrements of a wealthy Finnish family in 1763. *www.porvoonmuseo.fi; 50min by car or 1hr by train.*

03 **Nuuksio National Park**
The measure of success in Finland is neither fame nor fortune, but more quiet time in nature. If you haven't built your requisite Finnish cottage yet, head to Nuuksio, a microcosm of the Finnish landscape: quiet, unassuming, filled with forested paths and two dozen of Finland's 187,888 lakes. Several hotels and tour operators have saunas open for visitors. *www.nationalparks.fi/en/nuuksionp; 1hr by car or bus.*

04 **Fiskars Village**
Fiskars started as an ironworks company in the 1600s (you might have a pair of their scissors), but Fiskars Village has turned into an artists' colony. Dozens of working artisans sell jewellery, furniture, pottery, and other handcrafted designs. The village, shops and restaurants are spread throughout the company's historic buildings: old brick factories, the owner's manor house, and the 19th-century mill and foundry. *www.fiskarsvillage.fi; 1hr 10min by car or 2hr by train.*

05 **Hanko**
The population's 5% of Swedish-speaking Finns live mostly on the western coast and the semi-autonomous Åland Islands. One of the closest Swedish-speaking towns to Helsinki is Hanko (Swedish: Hangö), a cute-as-a-button beach resort town. Brightly coloured buildings and a boardwalk running along the scenic rocky coast make this a delightful day trip, especially in summer when the sandy beaches are filled with sunbathers thawing out after the long Finnish winter. *1hr 30min by car.*

──────TWO HOURS FROM──────

06 **Gegwen Husky Rides**
Do the happy puppies want to go for a run? Do they? Yes, they do! On your 2km ride (about 10–15 minutes) on a dog sled across wintry terrain, you won't be sure who's more excited about the exhilarating experience, you or the puppies. Post-mushing cuddles most definitely included. *www.gegwen.com; Christmas-Apr; 2hr by car.*

07 **Moomin World**
Got kids? Like warm-hearted entertainment that's good for the soul? This

---

THE LOCAL'S VIEW

'I love medieval Turku for a day escape (and its restaurants!) and Tallinn for the arts, architecture and handcrafts. Another smaller spot is Hämeenlinna (1hr 10min by car, 1hr 40min by bus or 1hr 30min by train), which lends itself perfectly to picturesque walks. The vast national park Aulanko (www.nationalparks.fi/en/aulanko) is a short stroll from the town centre. Plus, 13th-century Häme Castle is wonderful and has a fascinating adjacent prison museum. Pop into Piparkakkutalo for a romantic meal before strolling back to the bus or train.'

Nora Mäki, freelance PR consultant

© massimofusaro / Getty Images; © Jonathan Smith; © Minkimo / Alamy Stock Photo

waterfront amusement park is filled with all things Moomin, the hippo-like creatures that Tove Jansson created in her beloved book and television series. Greet Moomin characters in their world – in the Moomin garden, the witch's labyrinth and the five-storey Moomin House. Open daily in summer but check the website for off-season opening times. ***www.moominworld.fi; 2hr by car or 3hr by bus.***

08 Tampere
Here's how hard the 'Manchester of Finland' rocks: its claim to fame is death heavy metal, Gothic industrial warehouses and blood sausage. These days, Tampere hosts many a food festival and microbrewery. Try the *mustamakkara* (blood sausage) topped with lingonberry jam at the year-round indoor market hall, Kauppahalli. ***www.vanhakauppahalli.fi/en; 2hr by car or 2hr 30min by train.***

09 Turku
Once Finland's capital, Turku's medieval streets are now filled with stately historic buildings and shops frequented by the free-spirited city's artists and students. The imposing Turku Castle was built in 1280 by invading Swedish conquerors as a military fortress, but now hosts more tourists than invaders. Test out the innovative New Nordic restaurant scene at Kaskis, where dishes feature reindeer or foraged mushrooms; or at riverfront Tintå overlooking the cathedral. ***www.kaskis.fi/en, www.tinta.fi; 2hr by car or bus.***

——— THREE HOURS FROM ———

10 Tallinn, Estonia
Tallinn is filled with historic architecture, and stages a Christmas market which hosted its first Christmas tree display in 1441. In the Old Town, pick up traditional knits, handmade linens or juniper wood handicrafts. The Balti Jaam flea market is top choice for bargain antiques, clothes and an inexpensive lunch. Save half your euros by stocking up in the duty-free shops on the ferry back to Finland. ***3hr by ferry.***

© Suratwadee Rattanajarupak / Shutterstock

02

11 Seinäjoki Tango Festival

Argentina, step (dramatically) aside. Tango fever hit Finland in the 1960s; the melancholic melody appealing to the Finns' serious, somewhat earnest side. Since 1985, over 100,000 people have descended upon tiny Seinäjoki for the world's oldest tango festival, Tangomarkkinat. There are dozens of performances and dancing championships, and the winners of the popular singing contest 'Tango Royals' gain lifelong fame. *www.tangomarkkinat.fi; mid-July; 3hr by train.*

12 Kyrö Distillery

In a country known for vodka, the Kyrö distillery has made its name in gin and whiskey. Its rye gin, Napue, flavoured with juniper, locally foraged cranberries and birch, couldn't be more Finnish. Add some tonic water, fresh cranberries and a rosemary twig and you'll feel like you're swigging Finland in a glass. On Saturdays at 4pm, Kyrö opens to visitors with a tour, gin tasting and, on the first Saturday of the month, late-night live music. *www.kyrodistillery.com; 3hr by train.*

LÖYLY SPA & SAUNA

Löyly is Finland's version of Iceland's Blue Lagoon. Its name comes from the word for the steam rising from sauna stones, and this beautifully designed spa typifies the Finnish sauna culture of warm relaxation, quiet camaraderie and enjoyment of nature. Located in a Helsinki suburb, the spa-restaurant-beach complex (www. loylyhelsinki.fi; 30min by car or bus) is a full day-trip experience. Finnish saunas are usually gender-segregated and enjoyed in the buff, but here saunas are mixed, so make sure you bring a bathing suit.

Courtesy of Kyrö Distillery

EUROPE'S FINEST OFF-THE-BEATEN-TRACK WINERIES

Bordeaux, Champagne: so far, so predictable for European wine tourism. Call yourself an oenophile? Seek out the spittoons in these remoter regions.

Hush Heath Estate, England

It's the only English vineyard creating rosé sparkling wine (the award-winning Balfour Brut). Enjoy it from the panoramic tasting bar that opened in September 2018, overlooking 400 acres of Kent countryside and the Tudor estate dating back to 1503.
www.hushheath.com; 10min from Staplehurst train station by car.

Antinori Chianti Classico, Italy

An architecturally innovative setting that took 10 years to create? An art museum? A fabulous rooftop restaurant with views across the Chianti hills? *Sì* times three. Plus the Antinori family reputation: they've been making wine since 1385.
www.antinori.it/en/ tenuta/tenute-antinori/ antinori-nel-chianti-classico; 40min from Florence/Siena by car.

Domaine Chappuis, Switzerland

See why the Swiss mostly keep all their wine to themselves at this family estate (22nd generation of winemakers) in the Lavaux Vineyard Terraces overlooking Lake Geneva.
www.domainechappuis.ch/en; 20min from Lausanne by car.

Domaine de Cabasse, France

Renowned for its syrahs, this 33-acre vineyard in Provence has a restaurant, a pool and, importantly, a hotel: the perfect excuse to stay the night and devote yourself to some serious tasting and lingering walks among the lavender-lined paths and olive groves.
www.cabasse.fr/en; 1hr 30min/2hr 30min from Montpellier/Lyon by car.

Bodegas Tradición, Spain

Sherry's sexy again, have you heard? You need to visit this unique bodega in Jerez, which also houses Colección Joaquín Rivero, an art gallery of Spanish grand master paintings. The extra-aged sherries and brandies are critically acclaimed.
www.bodegastradicion.es/index.php/en; 1hr from Seville by car.

Tokaj Macik Winery, Slovakia

One of the smallest, most underrated wine regions in the world, the Slovak Tokaj area is worthy of your attention. This family-run, international-award-winning winery in Mala Trna also has a guesthouse; handy after overenthusiastic quaffing of the traditional Tokaj varietals.
www.tokajmacik.sk/en; 1hr from Košice airport by car.

Korta Katarina, Croatia

One of the few wineries that blends Plavac Mali from Dingač and Postup appellations, Korta Katarina has vineyards on Croatia's southern coast, overlooking the Adriatic. Since 2017, villas have been available for the ultimate seaside getaway.
www.kortakatarina. com/winery; 2hr from Dubrovnik by car.

Wine Art Estate, Greece

This small boutique winery near Drama in northern Greece has more than 100 international medals to its name. The unique microclimate (semi-continental rather than Mediterranean) and proximity to the Rodopi mountain range, which protects vineyards from the gusty northern winds, is part of the secret.
www.wineart.gr; 2hr from Thessaloniki by car.

Dr Pauly Bergweiler Wine Estate, Germany

Let the good doctor cure what ails you at this ultra-modern winery on Germany's Mosel River. Exceptional dry Rieslings from the steeply sloped vineyards are the main drawcard.
www.pauly-bergweiler. com; 2hr from Cologne by car.

Bodedga Artadi, Spain

This region of Spain might be renowned for Rioja, but owner and winemaker Juan Carlos López de Lacalle has turned his back on Rioja's traditional ageing methods at Artadi, focusing on single-vineyard origin, organic farming and respect for the terroir.
www.artadi.com/en; 1hr 20min from Bilbao by car.

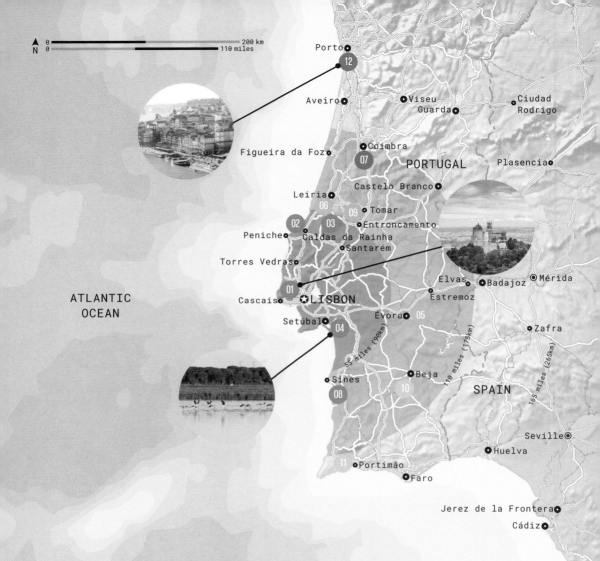

Porto○ 12

Aveiro○   ○Viseu
Guarda○   Ciudad
                 Rodrigo○

Figueira da Foz○   ○Coimbra
                   07   PORTUGAL   Plasencia○

                Castelo Branco○

Leiria○ 06
         09 ○Tomar
02   03 ○Entroncamento
Peniche○ Caldas da Rainha
         ○Santarém

Torres Vedras○

                                Elvas○ ○Badajoz ○Mérida
Cascais○ 01 ⊛LISBON   Estremoz○

Setúbal○ 04   ○Évora 05
                                          Zafra○

ATLANTIC
OCEAN

                   55 miles (99km)   110 miles (175km)   165 miles (265km)

         ○Sines ○Beja   SPAIN
         08   10

                              Seville◎

                      ○Huelva

11 ○Portimão
    ○Faro

              Jerez de la Frontera○
                      Cádiz○

LISBON

It's no secret that Lisbon is one of Europe's coolest cities, and now the rest of Portugal is becoming known for its enthralling architecture, landscape, history and gastronomy.

● ARTS & CULTURE   ● HISTORY   ● OUTDOORS   ● FOOD & DRINK   ● FESTIVALS & EVENTS   ● MUSIC & FILM

ONE HOUR FROM

01 Sintra

Allow as many hours as possible (or stay overnight if you can) at this popular day trip from Lisbon at the foot of the Sintra Mountains and revel in its forest of fairy-tale castles. There are many highlights, but the extravagant Palácio Nacional da Pena is beyond the craziest architectural dream, from the eclectic, luxurious interior to the pastel-palette exterior. *www.parquesdesintra. pt/en; 30min by car, 40min train from Lisbon's Rossio station.*

02 Óbidos Lagoon Wellness Retreat

This retreat centre just outside Óbidos really is 'in' nature: the wellness facilities include a fresh-air sauna, and massages are outdoors. You don't have to do the stand-up paddleboard yoga or go hiking, but the option is there, should you want it. Alternatives include swimming in the most massive salt-water swimming pool and chilling out in a Himalayan salt cave. *www.obidos-lagoon.com;1hr by car.*

03 Parque Natural das Serras de Aire e Candeeiros

Lace up your hiking boots because an array of fascinating caves and limestone cliffs await at this geological paradise of a natural park. Trails take you through contrasting landscapes – olive groves and hills dotted with gorse one minute, salt pans the next. But for most, it's all about the caves: 1500 of them. Start with the 683 steps down through the Mira de Aire caves (110m depth); there's a lift back up to the top. *www2.icnf.pt; 1hr by car.*

04 Reserva Natural do Estuário do Sado, Setúbal

A twitcher's paradise, this natural reserve that protects the Sado Estuary is home to some 250 species of bird, including flamingos, bee-eaters, white storks, owls and honey buzzards. It's a protected area of wetlands, and local tour guides are attuned to when birds might appear and where. In addition, bottlenose dolphins, a rarity in Europe, are often spotted on sightseeing cruises swimming in their natural environment. *www.natural.pt/portal/en; 1hr by car.*

05 Évora

A tiny, perfectly preserved medieval town encased in fortress walls, Évora delivers a sharp contrast with Portugal's popular coast. The flatness of the Alentejo landscape compared to Porto and Lisbon's steep hills is startling. The culinary focus also abruptly shifts from seafood to meat; the local black pork, in particular, is a highlight, as are the Azaruja sausages. *1h 30min by car or 1hr 45min by train from Lisbon's Gare do Oriente station.*

06 Mosteiro de Santa Maria da Vitória, Batalha

Almost a century in the making, this classic example of Gothic Portuguese architecture is a historically significant World Heritage Site: it was built to commemorate the 1385 Battle of Aljubarrota where 6500 Portuguese fought off Juan I of Castile's force of 30,000. Come prepared to strain your neck and 'ooh' and 'aah' at the intricacy of the stone carvings, the extravagant limestone exterior and the rare stained-glass windows. *www.mosteiro batalha.gov.pt; 1h 30min by car.*

PORTUGUESE PLATES

When dining out in Portugal, there's sometimes a bit of confusion when plates of food start appearing on your table before you've even ordered; not just baskets of bread, but often dishes of cheese, ham, olives, etc. No one has made a mistake: this is meant for you. But if you eat it, it will be added to your bill, which may explain why these plates often inexplicably seem to keep coming, looking more and more tempting. It's perfectly fine to ignore it all and just stick to ordering what you want from the menu.

──── TWO HOURS FROM ────

07 Coimbra

Coimbra tests your stamina with steep hills and even narrower, winding cobblestone lanes than in Lisbon, but the architecture and medieval charm is equally impressive. Don't miss Universidade de Coimbra (students still attend in long black capes) and its library, Biblioteca Joanina, one of the world's grandest. It even has a colony of live bats that feast on book-destroying insects at night! *1h 50min by train from Lisbon's Santa Apolónia station or 2hr by car.*

08 Cabeça da Cabra

This coastal retreat near Porto Covo in Alentejo is a scenic spot for surfing lessons. The owners restored an abandoned primary school into five studios. On a five-day retreat, all meals are provided, along with a daily surf lesson and morning yoga. Borrow a bike to cycle the miles of pristine beach or hike the Rota Vicentina trail. *www.cabecadacabra.com; 2hr by car.*

09 Covento de Cristo, Tomar

Get excited history buffs, this is Knights Templar territory. This magnificent former Roman Catholic monastery enclosed within 12th-century walls was the Knights' headquarters, founded in 1160 by Gualdim Pais, Grand Master of the Templars. There's some awesome architecture on display here, including the 16-sided Templar church, The Charola, its circular design supposedly allowing Knights to attend Mass on horseback. *www.conventocristo.gov.pt; 2hr by car.*

10 Herdade dos Grous, Albernoa

One sip of Portugal's wines and you'll realise that bedding down at a vineyard estate makes perfect sense. At this 1700-acre rural retreat, you can taste some of Alentejo's finest grape varieties, Trincadeira or Arangonez, in typically full-bodied wines. A horse tour (there stables here too) of the vineyards is recommended. There's also bikes, kayaks and paddleboats, the latter for use on the artificial lake. *www.herdade-dos-grous.com; 2hr by car.*

© Marina Denisova

08

11 Restaurante Caniço, Alvor

No, you're not in a James Bond film; to reach Restaurante Caniço you really do need to take an exclusive elevator down through a sandstone cliff, then walk through a tunnel. This ultimate culinary location specialises in seafood: you're right on the beach in a dining room carved out of a rock. It's hugely popular in summer (but mostly a 'word-of-mouth' destination), so book well ahead. *www.canicorestaurante.com; 12.30-4pm, 6.30-11pm; 2hr 30min by car.*

——— THREE HOURS FROM ———

12 Porto

Arrive in Porto in style, either by train at Sao Bento, one of the world's most beautiful stations – resplendent in azulejos – or by car across the glistening River Douro. Portugal's second city is high on cobbled-street charm, so sample the port, devour a Francesinha (see sidebar) and surrender your heart. *3hr by car or train from Lisbon's Santa Apolónia station.*

FRANCESINHA

Forget the *pastéis de nata* (those addictive, flaky Portuguese custard tarts) for a moment, there's a sandwich peculiar to Porto that is bewilderingly good in that so-wrong-it-feels-right unfathomable way. It's called the Francesinha and once you've had one you'll never forget it. It arrives looking like a giant melted cheese cube in gravy and, when carved open, you'll find layers of pork, then smoked sausage, bacon and beef between the bread. Beneath the cheesy encasing is also a fried egg, in case you felt this wasn't quite outrageous enough already. Best tried just once and reminisced about forever after.

© joyfull / Shutterstock

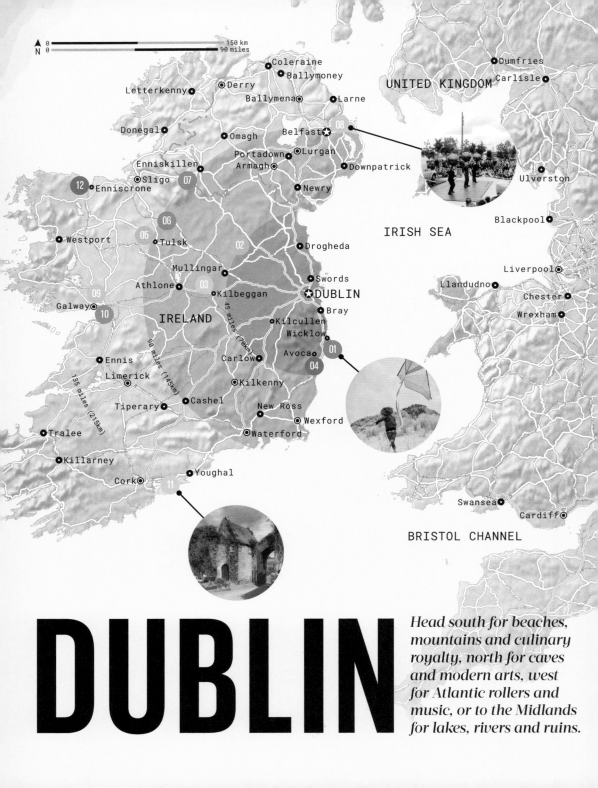

DUBLIN

Head south for beaches, mountains and culinary royalty, north for caves and modern arts, west for Atlantic rollers and music, or to the Midlands for lakes, rivers and ruins.

⬤ ARTS & CULTURE   ⬤ HISTORY   ⬤ OUTDOORS   ⬤ FOOD & DRINK   ⬤ FESTIVALS & EVENTS   ⬤ MUSIC & FILM

——— ONE HOUR FROM ———

01 Brittas Bay

Five kilometres of powdery sand, backed by undulating dunes, fringe the coast of Wicklow at Brittas Bay. For much of the year the Blue Flag beach is deserted and even in the height of summer, when half of Dublin descends on the place, it's easy to find your own patch of sand. Come early for a cinematic-style gallop on horseback along the waterline or try out a spot of gentle surfing. *www.visitwicklow.ie/item/brittas-bay-beach; 1hr 10min by car.*

02 Loughcrew Cairns

There's an unmistakable air of mystery to this 5000-year-old group of burial chambers and ritual monuments set across four hills in County Meath. Pick up a key from Loughcrew Gardens, climb the steep hill, make a wish from the Hag's Chair and enjoy the magnificent views. *www.heritageireland.ie/en/midlands-eastcoast/loughcrew; 1hr 10min by car.*

03 Kilbeggan Distillery

Unrepentantly traditional, this distillery was founded in 1757 on the banks of the River Brosna and rescued from collapse by the local community in time for its 250th anniversary. Today, it produces three Irish whiskeys and runs a guided tour to give you the low-down on its creaking waterwheel, massive steam engine and historic buildings, as well as a gargle of its malty magic. *www.kilbeggandistillery.com; 1hr 10min by car.*

04 Avoca

This tiny village set in a lush Wicklow valley is home to the Avoca Handweavers, a cottage industry established in 1723 which has evolved into a purveyor of the finest Irish style. Tour the original mill to see their vibrantly coloured rugs and blankets being produced, browse the shop for silky-soft throws and tastefully rustic ceramics, then feast on delicious salads, pies and pastries in the cafe. *www.avoca.com; 1hr 20min by car.*

——— TWO HOURS FROM ———

05 Cruchan Aí

An entrance to the underworld, seat of Iron Age warrior Queen Medbh and former capital of Connaught, Cruchan Aí is Europe's oldest and largest unexcavated royal complex, cunningly disguised as a cluster of bumpy fields. Head for the interpretive centre in Tulsk for help in deciphering the earthworks, burial mounds, ring forts, caves and ogham inscriptions that have lain undisturbed here for 3000 years. *www.rathcroghan.ie; 2hr by car.*

06 Shannon Blueway

A series of water- and land-based trails along the River Shannon between Lough Allen and Lough Ree, the Shannon Blueway offers a chance to explore a hidden side of Ireland, free from tour buses and souvenir stalls. Paddle, walk, cycle, stand-up paddleboard, swim, or stroll along the floating boardwalk across Acres Lake, and enjoy the genuinely warm welcome and old-school rural character along the way. *www.bluewaysireland.org/blueways/Shannon; 2hr by car.*

THE LOCAL'S VIEW

'If you like gardens but don't want to go too far, the walled garden in Marlay Park is always empty and a lovely place to go. Kilruddery House near Bray has gorgeous planting too and hosts a load of offbeat festivals, and if you're happy to travel a bit further, Mount Usher is a gem. Otherwise, I'd suggest cycling out of the city along the Grand Canal, it's really picturesque and it doesn't take long before you see herons, otters and all kinds of wildlife. It's a good flat, empty, cycle so everyone can do it.'

Oda Foyle, TV researcher & writer

Courtesy Belfast International Arts Festival; © monikahalinowska / Getty Images; © Design Pics Inc / Alamy Stock Photo

07 Marble Arch Caves

Carved out by a silent river beneath Cuilcagh Mountain in County Fermanagh, these 330-million-year-old caves comprise a network of winding passages and grand chambers dotted with subterranean waterfalls and sandy but sun-deprived beaches. A 75-minute tour includes a short boat trip along the river, when water levels allow, and a chance to explore a landscape far less well-known than the now-famous boardwalk above it. *www.marblearchcaves geopark.com; 2hr 10min by car.*

08 Belfast International Arts Festival

Held over a period of three weeks from mid-October to early November, Belfast's biggest arts festival offers theatre, dance, music, film and literature in a host of venues across the city. It's a great chance to combine a visit with world-class performances, workshops, talks, tours and open studio events, and see the city from an entirely different angle. *www.belfastinternationalartsfestival. com; 2hr 10min by train from Connolly Station.*

09 Galway Oyster Festival

Try your hand at shucking oysters or let someone else do the hard work and wash 'em down with champagne or stout. Established in 1854, the world's longest-running oyster festival is held on the last weekend in September. Galway bursts into life for this packed weekend of cooking demos, family fun, live music, tasting events and seafood trails. *www.galwayoysterfestival.com; late Sep; 2hr 20min by train from Heuston Station.*

10 Tig Coili, Galway

Crawl on over to this Galway institution, beloved for its warm welcome, great pints and nightly trad (traditional music) sessions. Neither trendy nor polished, the pub's Anaglypta-covered walls are laden with musical memorabilia, and it's often standing-room-only by nightfall. Come early if you want a seat and be prepared to fight your way to the bar. *www.tigcoiligalway.com; 2hr 20min by train from Heuston Station.*

© Ondrej Prochazka / Shutterstock

07

——— THREE HOURS FROM ———

11 **Ballymaloe**

The grande dame of Irish celebrity chefs, Myrtle Allen championed local, seasonal produce long before it hit the culinary headlines. Allen opened her first restaurant in her home over 50 years ago; today, her empire includes a cookery school, a restaurant, a hotel, a craft shop and gardens, all with the same down-to-earth values. *www.ballymaloe.com; 3hr by car.*

12 **Enniscrone seaweed baths**

Forget fancy day spas and head instead to the Kilcullen Seaweed Baths for a traditional homeopathic treatment, offered here since 1912. Little has changed since – the simple rooms, cedar steam cabinets, giant porcelain baths and original brass taps ooze nostalgia. Steam your pores, slither into the strangely velvety water and let the seaweed writhe about you, soften your skin and cure all manner of ills.
www.kilcullenseaweedbaths.net; 3hr by car.

09

© MariaKovaleva / Shutterstock

CAR-FREE TRIPS

If you haven't got much time, or your own wheels, there are plenty of great escapes right on Dublin's doorstep that are easily accessible by public transport. Hop on the DART (Dublin Area Rapid Transport) for a walk around Howth Head (30min by train from Connolly Station) or a clifftop ramble between Greystones and Bray (50min by train to Bray from Connolly Station). Take St Kevin's bus to Glendalough (1hr 30min) for early Christian history and mountain views, wander as far as Bull Island for birdwatching and seal spotting, kayak down the Liffey, or hop on a boat to Dalkey Island to see tern colonies and feral goats.

NORTH AMERICA

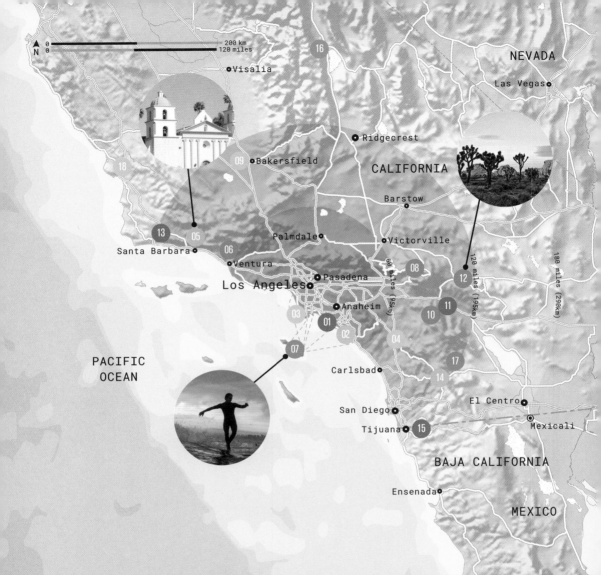

Outside LA, relaxed towns, arty events and unexpected food scenes exemplify the best of California's creative and mellow culture. And – whoa Nelly! – the nature: dramatic national parks, surreal film locations and meditative sunset spots.

LOS ANGELES

● ARTS & CULTURE　　● HISTORY　　● OUTDOORS　　● FOOD & DRINK　　● FESTIVALS & EVENTS　　● MUSIC & FILM

——— ONE HOUR FROM ———

01 Huntington Beach

Combine a visit to laid-back 'Surf City USA' with a slow comedy performance. After a day at the beach and the requisite lobster tacos, round things off at the Rec Room Comedy Club. Headliner friends of the comedian owner often try out new material here, which means longer, more in-depth sets. *www.recroomhb.com; 50min by car or 2hr by bus & train.*

02 Laguna Beach

The Pageant of the Masters event, an annual tableau vivant, started in 1932 and still has residents (who are really, really good at standing still) live-modelling famous paintings in front of a host of props. At the accompanying Festival of the Arts you can browse for artisan jewellery, crafts, paintings and everything in between, at this artsy seaside resort town. *www.foapom.com; Jul & Aug; 1hr by car or 2hr 10min by bus.*

03 Terranea Resort

In less than an hour from downtown (on a holiday Friday, maybe an hour and a half), you could be dining, golfing, hiking, sleeping or spa-ing at a luxurious cliffside resort. Terranea ticks off salt-water pools, haute cuisine, microbrews, tide pools for the kids, and activities ranging from falconry demonstrations to Mandala painting. On your way, stop by Rancho Palos Verdes' stunning Wayfarer's Chapel, a Swedenborgian glass church designed by Lloyd Wright (Frank's son). *www.terranea. com; 1hr by car or 2hr 10min by bus.*

04 Temecula

Leave Napa its overpriced, crowded tastings (and eight-hour drive from LA) and head to the nearby Temecula wineries instead. Walk through Old Town for a Wild West vibe, or towards Falkner Winery for a view over the valley (or ride one of the area's many hot-air balloons). In the decades since its inception, Temecula Valley wines have developed a surprising complexity; try the pioneer wineries of Calloway or Hart. *www.temeculawines.org; 1hr 30min by car.*

05 Santa Barbara

History meets art in this Spanish-style beachfront city that has it all: a mission, art museums and famous painters, world-class restaurants and a thriving university in nearby Isla Vista. Take a walking tour with the Architectural Foundation of Santa Barbara, ideally in May when the bright purple jacaranda trees are in bloom. Or celebrate the city's original Mexican, Native American and Spanish inhabitants at the annual Old Spanish Days Fiesta in the first week of August. *1h3 0min by car or 2hr 30min by train.*

06 Ojai

This mountain town has been working on its reputation as a centre of spiritual wellness for decades now. Wander through Libby Park's old oak trees or criss-cross one of the two dozen hiking trails, many with views of the scenic Ojai Valley. Upmarket hotels offer spas and wellness classes, or visit Meditation Mount for a citrus-tree-scented glimpse of Ojai's famous 'pink moment' sunsets. *1hr 30min by car or 3hr by bus.*

COACHELLA

From Beck to Daft Punk, Morrissey to Beyoncé, Coachella has grown in fame (and attendance) since it began in 1999. Officially known as the Coachella Valley Music and Arts Festival (www. coachella.com; two three-day weekends in mid-April; 2hr by car), the mega-festival attracts a quarter of a million concert-goers each year. The music and fashion set the trends for the following year, but there's also a thriving art and sculpture scene. Camping is now part of the experience, but come prepared with sunscreen and water – temperatures top 100°F (37.7°C). Showers are provided, thankfully.

07 Catalina Island

Here's the secret to a peaceful Catalina getaway: after the ferry docks in charming Avalon, head out to explore. Go on foot or with a tour, by bike, kayak or golf cart – there's a 20-year waiting list to own a car on the island. You'll have a chance of spotting Catalina's most famous Hollywood residents: bison. A 1924 film crew left a herd after shooting a western, and 150 now call Catalina home. *www.catalinachamber.com; 40min by car to ferry terminals in San Pedro & Long Beach, then 1hr ferry.*

08 Big Bear Lake

While most outsiders don't equate skiing with Southern California, the mountains northeast of LA are filled with a half-dozen world-class ski resorts. The twin resorts of Snow Summit and Bear Mountain in Big Bear serve up a combined 1000 acres of skiable terrain. Off the slopes, visit the nearby Big Bear Alpine Zoo: a wildlife rescue and rehabilitation centre for the area's grizzly bears, wolves, and eagles. *www.bigbearmountainresort.com; www.bigbearzoo.org; 1hr 50min by car.*

09 Bakersfield

File under 'Foodie Scene, Most Unexpected.' Basque shepherds and vintners helped settle America's breadbasket in Central California in the late 19th century. The community is still going strong, with a clutch of restaurants and a Basque festival each year over Memorial Day weekend. Taste traditional dishes, such as pickled tongue, for yourself in old-world restaurants like the

© Kevin Panizza / Getty Images

Noriega's, which still has a Basque pelota court. *www.noriegahotel.com; 2hr by car or 3hr by train.*

10 Idyllwild

Alpine cottages, secluded A-frames and cosy log cabins give this tiny mountain hamlet its fairy-tale feel. Nature and the arts battle for top billing, with performances at the famous residential Idyllwild Academy competing with rock climbing and hiking in Mount San Jacinto State Park. Mountain-biking trails criss-cross the foothills and backcountry, and nearby Lake Hemet is a haven for fishing and kayaking. *2hr by car.*

Palm Springs

While it started as a spring break destination, the Palm Springs desert communities now cater to retirees, shoppers, spa-lovers, and, as always, a thriving gay scene. There's also oasis hiking in nearby Desert Hot Springs or Aguas Caliente. Get a 6000m head start on the hike up Mount San Jacinto from the Palm Springs Aerial Tramway. *www.pstramway. com; 2hr by car or 3hr 30min by train.*

Joshua Tree National Park

Joshua Tree National Park, famous for its surreal rock outcroppings and spiny eponymous yuccas (technically a plant, not a tree), is now ground zero for Southern California's high desert art culture. The chilled vibe is evident at groovy art installations like the sculptures at the Noah Purifoy Foundation or quirky museums (crochet, anyone?). To explore your mind, stop by for a one-hour sound bath at the Integratron, a 'resonant tabernacle' that brings healing through crystal bowls. *www. nps.gov/jotr, www.joshuatree.guide/ artists; 2hr 20min by car.*

Solvang

Take early 1900s Denmark, add a giant helping of kitsch, then turn it up to an 11. Storybook architecture, mountains of pastries, Euro-themed shops – it's all here, in one surprisingly charming village. In December, Solvang hosts a traditional European Christmas market with live music and food stalls, and a community Christmas tree bonfire in early January. *2hr 30min by car.*

© Trinette Reed / Getty Images

14 Julian
Snuggle up in San Diego's crisp and cool mountain hamlet of Julian. After a mini-Gold Rush in the 1870s – the remnants still visible in historic buildings downtown – the town switched to cultivating apple orchards in 1907, and it's now practically the law to tuck into apple pie at a place like Julian Pie (www.julianpie.com) or Mom's Pie House (www.momspiesjulian.com). **2hr 40min by car.**

15 Tijuana
Every travel publication worth its weight in tequila salt has written about how Tijuana has recently morphed from a drinking haven to a sophisticated arts and culture scene, filled with boutique hotels, swanky bistros and shops, and the food truck collective, Telefónica Gastro Park. Tijuana's popular Cultural Center (CECUT) showcases the art and culture of northern Mexico, and Mercado el Popo offers the best of the area's handicrafts and food. **2hr 40min by car.**

16 Alabama Hills
Star Trek, Iron Man, Django Unchained, Gladiator, almost every single western ever – when Hollywood goes looking for a desolate, otherworldly landscape, they scout Alabama Hills, just west of Lone Pine. Its desert-like scrub and striking geological formations have been the backdrop for over 400 movie shoots. Just off Movie Flat Rd in the Alabama Hills Recreation Area is the Mobius Arch Loop Trail, where the eponymous arch folds scenically over Mt Whitney. **3hr by car.**

15

© Lindsay Lauckner Gundlock

16

© Mark Read

Anza Borrego Desert State Park

Los Angeles' light pollution is so extreme that anxious residents once called 911 during a blackout, to report a night-time sky disturbance (aka the Milky Way). Anza Borrego is so dark – especially in winter – it attracts astronomy fans from around the world. This unassuming state park also gets a lot of attention in the spring, when a multitude of wildflowers bloom.
www.parks.ca.gov/anzaborrego; 3hr by car.

San Luis Obispo

The oft-cited Happiest City in America is the Central California coast's college town of San Luis Obispo. SLO has laid-back vibes, a thriving food and bar scene and a farmers market so good Californians voted it #1 in the state (head to Higuera St on Thursday evenings). It's also surrounded by coastal charm, from the beach towns of Morro Bay and Los Osos to Hearst Castle and the quaint village of Cambria. *3hr 10min by car.*

LA TRAFFIC

LA traffic gets a bad rap. If you're visiting, know that in reality, well, yeah, it's pretty bad. However, carpools (either two or three occupants in the HOV 'diamond' lane) help bypass some of the traffic. In rush hour, some of the journey times listed might take twice as long, which is why we've included some spots close to LA that feel like a distant beach vacation: Terranea in Palos Verdes or Huntington Beach and Laguna Beach in Orange County. Plus, public transport is expanding. Comfy Amtrak trains run to Ventura/Oxnard, Bakersfield, San Luis Obispo and Santa Barbara.

© Shelly Rivoli / Alamy Stock Photo

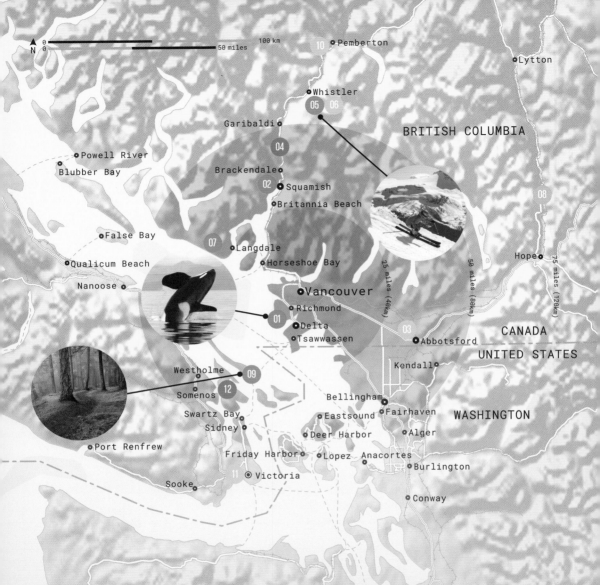

N
0
0
50 miles
100 km

10 Pemberton
Lytton

Whistler
05 06
Garibaldi
BRITISH COLUMBIA
04
Brackendale
02 Squamish
Britannia Beach

Powell River
Blubber Bay

False Bay
07 Langdale
08
Qualicum Beach
Horseshoe Bay
Hope
Nanoose

Vancouver
Richmond
01
Delta
Tsawwassen
03 Abbotsford
CANADA
UNITED STATES
Kendall

25 miles (40km)
50 miles (80km)
75 miles (120km)

Westholme 09
12
Somenos
Swartz Bay
Sidney
Bellingham
Eastsound Fairhaven
Deer Harbor
WASHINGTON
Alger

Port Renfrew
Friday Harbor
Lopez Anacortes
Burlington
11 Victoria
Sooke
Conway

*The natural drama of Canada's west coast makes outdoor getaways essential,
whether in the mountains, on nearby islands or in river canyons. Go whale
watching or skiing then explore indigenous culture and try local food and drink.*

VANCOUVER

● ARTS & CULTURE   ● HISTORY   ● OUTDOORS   ● FOOD & DRINK   ● FESTIVALS & EVENTS   ● MUSIC & FILM

——— ONE HOUR FROM ———

01 **Whale watching, Steveston**
Once the centre of British Columbia's salmon-canning industry, Steveston Village in the Vancouver suburb of Richmond is the departure point for three- to five-hour whale-watching tours. You'll search for orcas and humpbacks, plus sea lions, porpoises and other marine life that reside off the BC and Washington coasts.
www.seabreezeadventures.ca; Apr-Oct; 1hr from Waterfront Station by Canada Line, then bus from Richmond-Brighouse Station, or 1hr by whale-watch company shuttle.

02 **Squamish**
In this mountain-adventure hub, hike to the top of 'The Chief', a gigantic rock face that's also a magnet for climbers. Ride the Sea-to-Sky Gondola to a network of hiking trails, or challenge yourself on the Via Ferrata. Go sailing in Howe Sound, raft the white water on the Elaho-Squamish River, or join the windsurfers and kiteboarders soaring at Squamish Spit. *1hr by bus or car.*

03 **Fraser Valley**
Many farms in this agricultural valley east of Vancouver welcome visitors, particularly in summer and fall. Pick apples or blueberries, visit a cheesemaker, or find your way through a corn maze. Take in a flower festival, learn how honey is made, or try milking a cow. The valley has several small wineries too. Follow the self-guided Circle Farm Tour to find delicious experiences throughout the region. *www.circlefarmtour. com; 1hr 30min by car.*

———TWO HOURS FROM ———

04 **Sea-to-Sky Highway**
The Sea-to-Sky Highway winding between the ocean and mountains from Vancouver to Whistler isn't just a scenic drive. Kiosks at viewpoints en route tell you about the region's indigenous communities. In Whistler, on the traditional territory of two First Nations, the Squamish Lil'Wat Cultural Centre explores the history and culture of these indigenous peoples, while the Audain Art Museum showcases notable aboriginal art. *2hr by car.*

05 **Whistler**
Try bobsledding, ice fishing or dog sledding, or tackle the 200+ skiing and snowboard runs at Whistler Blackcomb, North America's largest winter sports resort. Whistler plays hard in the summer as well, whether you're hiking, mountain biking, ziplining, canoeing, sightseeing from a gondola that has the world's longest unsupported span, or taking a multimedia night walk through the old growth forest. *www.whistler.com; 2hr by bus or car.*

06 **Cornucopia Festival, Whistler**
Every November, Whistler draws food and drink lovers to the Cornucopia Festival for chef dinners, wine tastings, seminars and masterclasses, and gala parties. Come for a day, a weekend or more – the festival runs for 10 days – and while the ski season won't have started, you can balance your epicurean experiences with hiking, yoga or just chilling in the mountains.
www.whistlercornucopia.com; Nov; 2hr by bus or car.

THE LOCALS' VIEW

'I go hiking in Lynn Valley and Lynn Headwaters Park,' says Brad Miller, chef-owner at The Red Wagon and Wagon Rouge restaurants. 'I feel I'm far from the city in about 20 minutes.'

'We love the ferry to the Sunshine Coast. We look for family-friendly B&Bs with access to the beach.' Katharine Manson, communications manager

'You can't beat hiking "The Chief" on a sunny day. The reward comes afterwards – sampling beers at a microbrewery nearby.' Leah Heneghan, director of Vancouer Craft Beer Week

© stockstudioX / Getty Images; © Michael Wheatley / Getty Images; © MarkMalleson / Alamy Stock Photo; © MarkMalleson / Getty Images

07 Sunshine Coast

Hop on a ferry to the Sunshine Coast, a peninsula of rocky beaches and small towns accessible only by water. Hike to Skookumchuck Narrows, where tides churn up unusually large rapids twice a day (fuel your hike with freshly baked cinnamon rolls from a trailhead bakery), or cruise through the fjords between towering granite cliffs to Chatterbox Falls, which plunge into the sea. *2hr by car or bus and ferry.*

08 Fraser Canyon

In the 1850s, prospectors travelled to this deep gorge along the Fraser River, searching for gold. The Fraser Canyon, which the Trans-Canada Highway now follows between Hope and Cache Creek, had another boom when Canada's transcontinental railroad crossed the region later that century. Learn more at the Yale Historic Site, then descend into the canyon on the Hell's Gate Airtram or go white-water rafting along the river. *2hr by car.*

09 Island Escape, Galiano Island

For a quick island escape, board the one-hour ferry to quiet Galiano, one of BC's Gulf Islands. Explore the beach at Montague Harbour, hike up the tree-lined trails on Mount Galiano, then head to farm-to-table dining destination, Pilgrimme. To stay longer, book an island cottage or settle into the oceanside Galiano Inn & Spa. *www.pilgrimme.ca, www.galianoinn.com; 2hr by car & ferry.*

10 Slow Food Cycle Sunday, Pemberton

Explore the farms and orchards in the Pemberton Valley, an agricultural area between the coastal mountains north of Whistler. For one Sunday in August, the local chapter of the Slow Food organization hosts a bike ride along Pemberton's country roads. Stop at farms and food stands, and enjoy an outdoor BBQ, as you pedal through this rural region. *www.slowfoodcycle sunday.com; 2hr 30min by car.*

———— THREE HOURS FROM ————

(11) Victoria

On Vancouver Island, the city of Victoria has long been a destination for traditional afternoon tea, particularly at the venerable harbourfront Fairmont Empress Hotel. These days, the craft beer, cider and spirits scene also draws sippers to local breweries and pubs. To pedal between microbreweries, join bicycle tour company, The Pedaler, for a three-hour 'Hoppy Hour Ride'. *www.thepedaler.ca; 3hr by ferry or bus-ferry combination, 40min by float plane or helicopter.*

(12) Salt Spring Island Studio Tours

More than two dozen artists and craftspeople open their studios to visitors on Salt Spring Island, the largest of the Southern Gulf Islands, off BC's coast. On this year-round, self-guided tour, you might chat with potters, jewellery designers and other artisans about their work. Drop in at a cheesemaker, cidery or microbrewery for refreshments. *www.saltspringstudiotour.com; 3hr by car and ferry.*

VANCOUVER FERRIES

BC Ferries (*www. bcferries.com*) operates two terminals in the Vancouver area, one south of the city at Tsawwassen, where boats to Victoria and the Southern Gulf Islands depart, and the other at Horseshoe Bay, for the Sunshine Coast. BC Ferries transport foot passengers, bicycles and cars. If you're taking a vehicle, it's a good idea to reserve in advance, particularly in the summer and holidays. A private ferry service, V2V Vacations (*www. v2vvacations. com*), runs a convenient but more expensive passenger-only catamaran from Vancouver and Victoria, mid-March to mid-October.

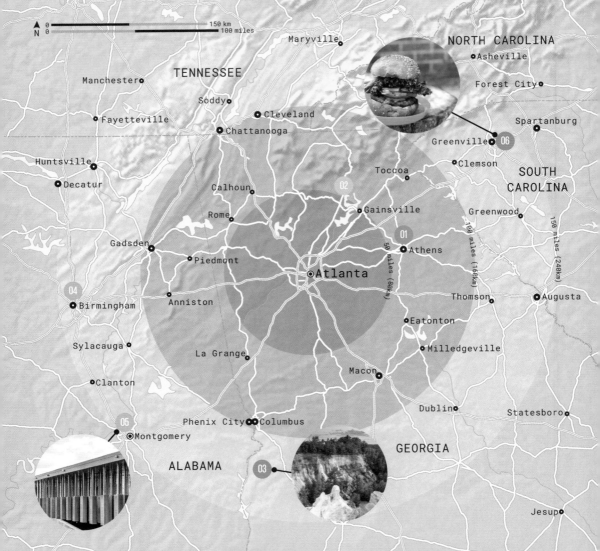

You'll never run out of things to do in this thriving pocket of the South. Hear avant-garde sounds, hike an impressive canyon and sip award-winning wines.

ATLANTA

● ARTS & CULTURE ○ HISTORY ○ OUTDOORS ○ FOOD & DRINK ○ FESTIVALS & EVENTS ○ MUSIC & FILM

———— ONE HOUR FROM ————

01 **40 Watt Club, Athens, Georgia**
This funky college town, home to the University of Georgia, launched the careers of REM, the B-52's and Widespread Panic, and its music scene is still lit. The 40 Watt Club has been the centre of Athens' musical universe for decades and hosts big names and regional stars. *www.40watt.com; 1hr 40min by car.*

02 **North Georgia Wine Country**
The Appalachian foothills north of Atlanta are now a recognised viticultural region, thanks to the ultra-quaffable wines produced here. There are more than a dozen wineries to choose from, many with breathtaking mountain views, live music or patios that are perfect for whiling away an afternoon. Pull up a stool at one of Dahlonega's tasting rooms to sample the best of the region. *www.dahlonega.org/see-do/wineries-a-vineyards; 2hr by car.*

03 **Providence Canyon State Park, Georgia**
You might not expect Grand Canyon-style landscapes within an easy drive of Atlanta, but that's what you'll find at Providence Canyon. The 1000-acre park has hiking trails galore amid its trademark gullies, with cliffs of layered sediment in a rainbow of colours. Stop for a bite and a riverside stroll in Columbus, on the Chattahoochee River. *www.gastateparks.org/providencecanyon; 2hr 20min by car.*

04 **Birmingham, Alabama**
Atlanta is full of Civil Rights sights, but many pivotal moments in this ongoing struggle happened in nearby Birmingham. The downtown Civil Rights Heritage Trail begins at Kelly Ingram Park – where you can see powerful statues commemorating police violence against peaceful protestors – and includes the Sixteenth Street Baptist Church. Be sure to stop at the Civil Rights Institute to add context to your walk. *http://heritagetrail.birminghamal.gov; 2hr 20min by car.*

05 **The National Memorial for Peace & Justice, Montgomery, Alabama**
One of the country's most important sights is just a short drive along I-85. The National Memorial for Peace and Justice is the first-ever tribute to the more than 4000 known victims of lynching in the United States. The 800 majestic monuments, one for each county where a lynching took place, make the scale of terror obvious, and the nearby Legacy Museum makes the connection between past and present clear. *www.eji.org/national-lynching-memorial; 2hr 20min by car.*

———— THREE HOURS FROM ————

06 **GHS Swamp Rabbit Trail, South Carolina**
Rails to Trails projects have produced beloved recreational opportunities across the US, and this 19-mile trail that runs from charming Greenville to quaint Traveler's Rest is one of the best. Joggers, cyclists and families can be found enjoying this relatively flat, shady greenway – stop at Swamp Rabbit Cafe & Bistro (www.swamprabbitcafe.com) to grab a picnic. *www.greenvillerec.com/ghs-swamp-rabbit-trail; 2hr 40min by car.*

PASAQUAN

Fans of unusual photo ops and outsider art shouldn't miss this unique attraction near Buena Vista (https://pasaquan.columbusstate.edu; 2hr 30min by car). After having visions in which he was chosen by 'people of the future' to depict their culture of peace and love, self-taught artist Eddie Owens Martin (1908–1986) turned his mother's 19th-century farmhouse into a psychedelic wonderland over the course of three decades. The site – which includes six buildings – is an explosive, rainbow-hued fusion of African, pre-Columbian Mexican and Native American motifs.

NORTH AMERICA'S CULTURE & HISTORY HUBS

Crystal Bridges Museum of American Art, Bentonville, Arkansas
Amid a lush Ozark forest, Crystal Bridges is an architectural wonder filled with works that take visitors through the development of American art. *www.crystalbridges.org; 3hr from Little Rock by car.*

Museum of the Rockies, Bozeman, Montana
Visitors will find a massive collection of fossils at this off-the-beaten-path museum, including one of the most complete T. rex skeletons ever discovered. *www.museumoftherockies.org; 2hr from Helena by car.*

Lost City Museum, Overton, Nevada
Learn about the lives of the Ancestral Puebloans through reconstructed homes and the artefacts that were saved as this desert land developed. *www.nvculture.org/lostcitymuseum; 1hr from Las Vegas by car.*

Mark Twain Boyhood Home and Museum, Hannibal, Missouri
Before he was Mark Twain, Samuel Langhorne Clemens grew up in this home in Hannibal, a town that would inspire the setting of *The Adventures of Huckleberry Finn*. *www.marktwainmuseum.org; 2hr from St Louis by car.*

Marfa, Texas
Packed with sculptures, installations, galleries and working artists, the tiny town of Marfa is a beacon of culture in a sparse desert landscape. *www.visitmarfa.com; 3hr from El Paso by car.*

Wander away from the big city to unearth some of the best art, culture and history on the continent in the most unexpected places.

Georgia O'Keefe Museum, Santa Fe, New Mexico

Arriving at this museum after a trip through the dramatic landscapes of the Southwest brings a fresh perspective to O'Keefe's art and her lasting influence on American modernism. *www.okeeffemuseum. org; 1hr from Albuquerque by car.*

Hunter Museum of American Art, Chattanooga, Tennessee

Standing imposingly on the banks of the Tennessee River, a historic mansion and its modern extensions hold a wealth of American art from the colonial period to the present. *www.huntermuseum. org; 2hr from Nashville by car.*

Glenstone Museum, Potomac, Maryland

Walking the pathways through woodlands and meadows is just the first step to exploring this centre of modern art, where visitors are encouraged to contemplate the collection, architecture and the beauty of nature. *www.glenstone.org; 1hr from Washington, DC by car.*

Blackfoot Crossing Historical Park, Alberta, Canada

Rising up from the open prairie, this cultural and historic site on a First Nations reserve takes visitors on an authentic journey into the history and traditions of the Blackfoot people. *www.blackfootcrossing.ca; 1hr from Calgary, Alberta by car.*

Ernest Hemingway Home and Museum, Key West, Florida

Hemingway's peripatetic life was spent around the world, but his Key West home – now filled with six-toed cats descended from his own – was where he wrote some of his best-known works. *www.hemingwayhome. com; 3hr 30min from Miami by car.*

The bayous, lakes and rivers near New Orleans brim with getaways that embody
Louisiana culture. Hit a road and pick a point on the compass. Soon, you'll be
immersed in the state's complex history, lively arts scene, Cajun culture and more.

NEW ORLEANS

● ARTS & CULTURE ● HISTORY ● OUTDOORS ● FOOD & DRINK ● FESTIVALS & EVENTS ● MUSIC & FILM

──── ONE HOUR FROM ────

01 Barataria Preserve, Jean Lafitte National Historical Park

This section of the Jean Lafitte National Historical Park and Preserve, south of New Orleans near the town of Marrero (and Crown Point), provides the easiest access to the encircling dense swamplands. The eight miles of boardwalk trails offer a stunning exploration through the fecund, thriving swamp, home to alligators, nutrias (basically big, invasive river rats), tree frogs and myriad species of birds. *www.nps.gov/jela/ barataria-preserve; 30min by car.*

02 Abita Brewery, Abita Springs

The bucolic village of Abita Springs on Lake Pontchartrain's north shore was popular in the late 19th century for its curative waters. Today spring water still flows from the town fountain, but, more importantly, beer bubbles from the Abita Brewery, one of America's largest craft beer producers. Buckets of seafood lead the brewery's pub menu; if it's springtime, try the strawberry seasonal. *www.abitabrewpub.com; 1hr by car.*

03 Laura Plantation, Vacherie

Discover Louisiana's turbulent past at Laura Plantation, a restored 19th-century estate on the west bank of the Mississippi River. It was run by four Creole women, most notably Laura, for whom it is named. A tour teases out the distinctions between Creole, Anglo, free and enslaved African Americans. Culturally and architecturally, this Creole mansion is strikingly distinct from other plantations. *www.lauraplantation.com; 1hr by car.*

04 Baton Rouge

Baton Rouge, Louisiana's capital, is home to a thriving arts community, largely thanks to Louisiana State University. The LSU Museum of Art, within the clean, geometric lines of the Shaw Center for the Arts, holds a permanent collection of over 5000 works and also hosts touring exhibitions. There are lots more galleries on and off campus, many affiliated with the university and its myriad arts courses. *www.lsumoa.org; 1hr 20min by car.*

──── TWO HOURS FROM ────

05 Shrimp Festival, Delcambre

Boiled shrimp, fried shrimp, grilled shrimp, shrimp salad, and pretty much any other shrimp dish you can think of, feature at this five-day festival in the rural fishing enclave of Delcambre, in western Louisiana's bayou country. The town's docks are lined with the local shrimp fleet while everyone celebrates. The fun includes cooking contests, live music and carnival rides. Don't miss the spectacular firefighter water battles. *www.shrimpfestival.net; mid-Aug; 2hr 10min by car.*

──── THREE HOURS FROM ────

06 Prairie Acadian Cultural Center, Eunice

Experience rural life and Cajun culture at this museum. The best time to visit is on a Saturday before noon when the Savoy Music Center (located in an old accordion factory) hosts a Cajun music jam session. Later in the day there are cooking demos and tastings. *www.nps.gov/jela/prairie-acadian-cultural-center-eunice; 2hr 40min by car.*

THE LOCAL'S VIEW

'I like to head across Lake Pontchartrain by car to the lush wetlands of Fontainebleau State Park. After a walk amid towering cypress trees, with occasional 'gator sightings, I visit the nearby town of Mandeville. Its quirkier sights include St Francis Sanctuary – a winning combination of thrift store and feline haven. The day ends with a meal at the elegant Trey Yuen. It's surrounded by koi ponds and manicured gardens, and serves the best Chinese cooking for miles around.'

Regis St Louis, author

© Richard Cummins / Alamy Stock Photo; © John Elk ; © Kris Davidson

NEW YORK

150 km
100 miles

06

Lebanon

11

12

07

Portland◎

NEW
HAMPSHIRE

Claremont◎

Saratoga
Springs
◎

VERMONT

04

◎Concord

50 miles (80km)
◎Manchester

Schenectady◎

Brattleboro◎

◎Milford

◎Troy
Albany◎

02

◎Lawrence

◎Leominster

10   08
◎Lenox
MASSACHUSETTS

◎Hudson

◎Boston

◎Worcester

05

◎Holyoke

Provincetown

◎Woonsocket

◎Brockton

◎Middleborough

01

Newburgh◎

◎Danbury

Norwich◎

◎Hartford

CONNECTICUT

New London◎

◎Providence

RHODE
ISLAND

03

◎Newport

09

◎Oak Bluffs

◎New Haven

Norwalk◎

◎Bridgeport

◎Mt Vernon

◎New York City

ATLANTIC
OCEAN

*Almost half of New England is within easy distance of Boston, from the cow-
dotted hills of Vermont and the arty villages of western Massachusetts to the
tumbling grey seas of Maine. Bus and ferry services make it easier still.*

BOSTON

● ARTS & CULTURE   ● HISTORY   ● OUTDOORS   ● FOOD & DRINK   ● FESTIVALS & EVENTS   ● MUSIC & FILM

——— ONE HOUR FROM ———

01 Providence, Rhode Island
The capital of America's smallest state offers a quirky good time. Brown University and the Rhode Island School of Design (RISD) give the city an arty, youthful vibe, with excellent coffee shops, dive bars, used bookstores and indie theatres. Hit the RISD Museum for art from classical Greek to Andy Warhol, and Waterplace Park for art installations and summer concerts.
1hr by car or 30min by train from South Station.

02 Western Massachusetts
It wouldn't be fall in New England without hot apple cider and cinnamon-spiked cider donuts, both harvest traditions in the apple orchards of western Mass. You can create a whole day of autumn memories at fourth-generation family-run Red Apple Farm in Phillipston. Pick from 50 varieties of apple, munch donuts and fudge, visit the goats, take a hayride, then sip an apple beer in the seasonal 'Brew Barn'.
www.redapplefarm.com; 1hr 10min by car.

03 Mystic, Connecticut
This seaside village was once one of America's most important whaling and shipbuilding centres, and Mystic now trades on its salty history. Visit the Charles W. Morgan, the world's oldest wooden whaling ship, at the 40-acre Mystic Seaport Museum, wander the historic riverfront downtown, take a schooner cruise, then tuck into clam chowder at the Captain Daniel Packer Inne, dating from 1756. *1hr 30min by car or by train from South Station.*

———TWO HOURS FROM———

04 Brattleboro, Vermont
With a population of just 12,000, Brattleboro packs an outsized cultural punch. The Brattleboro Museum & Arts Center has been showing cutting-edge contemporary work for nearly 50 years, while the galleries of winsome downtown are a must-see for anyone hoping to fill a blank wall. Come in winter for the charming Winter Carnival, or in summer for the Strolling of the Heifers (yes, a parade of cows). *2hr by car.*

05 Provincetown, Cape Cod, Massachusetts
On the tip of crab-claw-shaped Cape Cod lies Provincetown, often proclaimed the most gay place in America. There are apparently more same-sex couples here than anywhere in the country, and each summer the irresistible resort town explodes with LGBTQ travellers from across the world. Swim (nude, if you dare!) at chilly Herring Cove Beach, drop in on the afternoon tea dance at the Boatslip, then catch a drag revue at the Crown and Anchor. *2hr by car.*

06 Cannon Mountain, New Hampshire
Deep in New Hampshire's White Mountains, this long-time state-run ski resort offers slopes for all levels. It's got the highest summit and longest vertical drop in the state, so thrill seekers won't be disappointed. The powder action starts in November and continues to mid-spring. For non-skiers, there's a fun aerial tramway plus plenty of hiking and fishing, not to mention après-ski pub grub and beer in the nearby villages. *2hr by car.*

<u>LEAF PEEPING</u>

Each autumn, New England's fiery foliage becomes a destination in itself. But where to head? For the top fall scenic drives, try New Hampshire's Kancamagus Highway, winding 35 miles through the White Mountains, or Vermont's rustic Route 100. Towns with the best foliage photo ops include Stowe, Vermont, at the foot of the state's highest peak, or New Marlborough, in the Massachusetts Berkshires. Hole up at the grand Omni Mount Washington Resort (*www.omnihotels.com/hotels*) or the Lodge at Moosehead Lake (*www.lodgeatmooseheadlake.com*), overlooking the russet hills of inland Maine.

© Jared Alden / Getty Images. Courtesy of Eventide oyster Co.; © gregobagel / Getty Images

07 Portland, Maine

Stroll from the Victorian sea captain's mansions of the West End to the cobblestones of the Old Port in search of foodie Portland's best bites. Seafood is, of course, always top-notch: suck down robustly briny Pemaquid oysters at edgy Eventide Oyster Co, crack open steaming whole lobsters at the Portland Lobster company, or gobble unagi rolls made with local eel at Miyake. *2hr by car or 2hr 30min by train from North Station.*

08 Tanglewood Music Festival, Massachusetts

The Boston Symphony Orchestra has had a summer residency at this bucolic estate in the Berkshire Hills for more than 80 years, drawing music-lovers from far and wide. Come for classical, jazz, chamber music, pop, musical theatre and more, with performances on Tanglewood's wide green lawns and in its star architect-designed concert halls. Pre-concert luxe picnics (champagne, smoked salmon, pâté) are de rigueur, so come prepared. *Jun-Sep; 2hr 10min by car.*

09 Oak Bluffs, Martha's Vineyard, Massachusetts

The town of Oak Bluffs on elite Martha's Vineyard, only accessible by air or ferry, has been a favourite destination of well-heeled African-American families for more than a century. Hit the African-American Heritage Trail, with 27 illuminating sites, from the vine-tangled Eastville Cemetery to the island's first African-American-run guesthouse. Then cool off at The Inkwell beach, whose once-pejorative name is now embraced.

mvafricanamericanheritagetrail.org; *2hr 30min by car & ferry or bus & ferry.*

10 Hudson River Valley, New York

Cross into upstate New York and fall back hundreds of years. The villages along the Hudson River are chock-a-block with antiques shops trading in furniture, art and curios rummaged from the area's old Dutch barns, Industrial Revolution factories and crumbling captains' mansions. Try Hudson for its sheer number of longstanding shops (mid-century modern is a speciality), Kingston for arty, well-curated wares and Beacon for its fun seasonal flea market. *2hr 30min by car.*

—— THREE HOURS FROM ——

11 Damariscotta, Maine

Drive up Maine's craggy Midcoast to the postcard-pretty village of Damariscotta.

© James Kirkikis / Shutterstock

Get classic photo ops at Pemaquid Point Lighthouse, guiding ships for 150 years from atop strikingly striated granite cliffs, and gawp at the massive oyster shell midden left behind by the native Abenaki people. Then spend the afternoon porpoise-spotting from a kayak on icy Muscongus Bay before heading back to the urban grind. *2hr 40min by car.*

Maine Lobster Festival
Oh, the iconic lobster! Once considered a poor man's meal, now such a delicacy it draws visitors to Maine just to taste its purest form. Each summer the village of Rockland serves 20,000lb of the crustacean in every imaginable guise – lobster rolls, lobster mac 'n' cheese, lobster ice cream – at its 71-year-old Lobster Festival. Expect carnival rides, concerts, cooking demos and the coronation of the Maine Sea Goddess. *www.mainelobsterfestival.com; late Jul to early Aug; 3hr 10min by car.*

BUSES, TRAINS & FERRIES

Though many New England getaways demand a car, there are plenty of public transportation workarounds, especially in high season. The Seastreak ferry to Martha's Vineyard now has a bus service from Boston to the ferry terminal in New Bedford. Boston Harbor Cruises offers a summer-only service from Boston to Provincetown in just 90 minutes. Concord Coach Lines connects Boston with Maine and New Hampshire by bus, though many of its routes take significantly longer than driving. Amtrak's Downeaster service goes from Boston through Midcoast Maine.

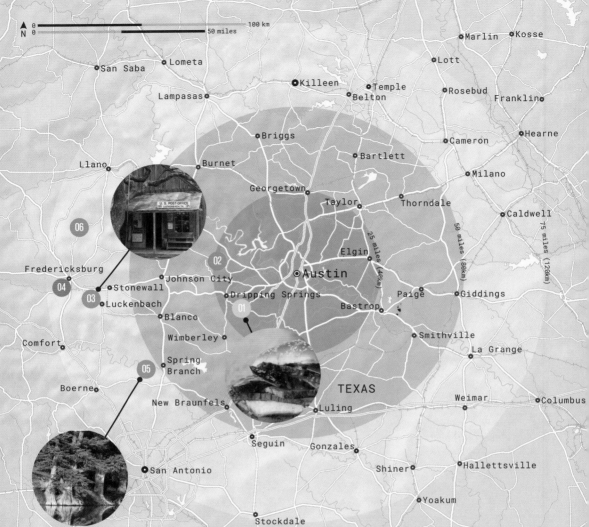

100 km
50 miles

N

Kosse
Marlin
Lott
Rosebud
Franklin
Hearne
Cameron
Milano
Caldwell
Killeen
Temple
Belton
Bartlett
Briggs
San Saba
Lometa
Lampasas
Llano
Burnet
Georgetown
Taylor
Thorndale
Giddings

06

Fredericksburg
04
03
Stonewall
Luckenbach
Johnson City
02
Austin
Elgin
Paige
Blanco
Dripping Springs
Bastrop
Comfort
Wimberley
01
Smithville
Spring Branch
05
Boerne
New Braunfels
TEXAS
Luling
La Grange
Weimar
Columbus
San Antonio
Seguin
Gonzales
Shiner
Hallettsville
Stockdale
Yoakum

25 miles (40km)
50 miles (80km)
75 miles (120km)

Spend your days tucking into world-class barbecue or exploring enchanting natural spaces. Come evening, head to outlaw country hideaways tucked into picturesque countryside, settled by the region's first German immigrants.

AUSTIN

ARTS & CULTURE   HISTORY   OUTDOORS   FOOD & DRINK   FESTIVALS & EVENTS   MUSIC & FILM

——— ONE HOUR FROM ———

01 Salt Lick BBQ, Driftwood
Head to the tiny town of Driftwood for lazy afternoons of world-class food and 'fun for the whole family' yard games at one of Central Texas' most popular barbecue joints. Large groups with even larger appetites can reserve all-you-can-eat 'Family Style' portions, but mere mortals opt for the meat plate combos of brisket; pork; beef and bison ribs; pulled pork; sausage and turkey on offer from the open-pit barbecues inside. *www. saltlickbbq.com; 30min by car.*

02 Hamilton Pool Preserve & Reimers Ranch Park, Dripping Springs
Hamilton Pool, a natural grotto fed by a 50ft waterfall, is a well-loved swimming hole. Make time for the nearby Reimers' Ranch Park, where 18 miles of trails and three miles of riverfront bluffs attract bikers and climbers, but not nearly as many as the swimmers congregating at Hamilton Pool, despite being only a mile away. *https://parks.traviscountytx.gov/ find-a-park/hamilton-pool; 50min by car.*

03 Luckenbach
Put on the map by Jerry Jeff Walker and thrust to national fame by Willie Nelson and Waylon Jennings, the unincorporated community of Luckenbach may have moved on from the strictly 'outlaw country' vibe, but it's still very much the heart of Texas country. There's rollicking live music every weekend and visitors are always welcome. *www. luckenbachtexas.com; 1hr 20min by car.*

04 Fredericksburg
With a wealth of events, wineries and in-town attractions, it's often hard to decide how to best spend a day in Fredericksburg. It was settled by some of Texas' first German immigrant families, and the European frontier ethos shines through in the architecture and history of the town itself. Further afield, vineyard tours are a hit with groups on weekend trips from Austin. *1hr 30min by car.*

05 Guadalupe River
There's no better respite from the Central Texas summer than jumping in the water, and few places could beat the Guadalupe River; specifically, drifting down its course on an inner tube. Head to Guadalupe River State Park for a family-friendly float (plus campsites and hiking), or look for local private operators that offer a more party-focused experience on the river. *www.tpwd.texas.gov/state-parks/ guadalupe-river; 1hr 30min by car.*

——— TWO HOURS FROM ———

06 Enchanted Rock State Natural Area, Fredericksburg
The 425ft high pink granite dome of Enchanted Rock towers over the surrounding Central Texas hills. The popular Summit Hike tracks past vernal pools and rock fissures to panoramic views from the top. Queues form at the State Park gate as early as 8am on busy weekends, but campers with confirmed reservations cruise straight on through and into the park. *www. tpwd.texas.gov/state-parks/enchanted-rock; 1hr 40min by car.*

THE LOCAL'S VIEW

'My favourite escape starts with "Old 300 BBQ" in the sleepy town of Blanco. The brisket is excellent and I love the wall map where diners mark what corner of the world they come from. After filling up on barbecue, I take the backroads west to Luckenbach and enjoy sitting under the large live oaks and listening to whatever musician happens to be playing. It's different every time, depending on the music and size of the crowd, but a great way to spend a few hours relaxing in the beautiful hill country on a Saturday afternoon.'

Tonie Dees, para-professional

Hamilton Pool Preserve, 50 minutes
from Austin.

NORTH AMERICAN W

Serious animal-spotting doesn't have to mean week-long safaris. Some of the finest sites for wildlife-watching are within a bunny hop of North American cities.

ILDLIFE GETAWAYS

Puffin Cruises, Maine
In summer, hop aboard a boat from Boothbay Harbor to Eastern Egg Rock to spy colonies of nesting puffins. The slow-moving black-and-white birds, with their waddly feet and forlorn eyes, make top photo subjects. *www.mainepuffin.com; $41; 1hr from Portland by car.*

Point Reyes Safari, California
Let a professional wildlife photographer guide you through the grasslands, pine forest and rocky beaches of Marin County's Point Reyes National Seashore, home to bobcats, elephant seals, raptors, whales, elk and more. *www.pointreyessafaris.com; half/full day for 3 people $595/$795; 1hr from San Francisco by car.*

Prairie Animals Driving Tour, Kentucky & Tennessee
The narrow strip of land between Lake Barkley and Kentucky Lake is a vast recreation area, including a 700-acre grazing prairie for elk and bison; visit them by driving tour along a loop road. *www.landbetween-thelakes.us; $5 per car; 2hr from Nashville by car.*

San Juan Island Orcas, Washington
Killer whales mob the Pacific Northwest's San Juan archipelago to gobble up salmon. Spy them from the deck of an orca cruise or the window of a seaplane. *www.sanjuansafaris.com; $100/person; 3hr from Seattle by car.*

Moose-Watching, La Forêt Montmorency, Québec
Catch moose grazing silently in the chilly dusk in this Canadian boreal forest, also home to black bears, timber wolves, beavers and more. In the September and October rutting season, you can hear antlers clacking from miles away. *www.foretmontmorency.ca/en; 1hr from Québec City by car.*

Elk- & Black Bear-Spotting, Great Smoky Mountains National Park, North Carolina & Tennessee
You may catch grazing elk or berry-picking black bears almost anywhere in the Smokies, but elk seem to favour the Cataloochee Valley, while bears prefer to wander among the historic cabins of Cades Cove. *www.nps.gov/grsm/index.htm; 3hr from Atlanta by car.*

Seals & Bison, Catalina Island, California
Just 22 miles southwest of Los Angeles lies Santa Catalina Island, whose rocky hills teem with deer, foxes and American bison. Just offshore swish leopard sharks, seals and the occasional great white. *1hr by ferry from Long Beach.*

Everglades Gators, Florida
A kayak tour is one of the best ways to wildlife-watch in this endless 'river of grass'. In addition to the ubiquitous gators, you might see dolphins, manatees, sea turtles and bald eagles. *www.evergladesarea-tours.com; $180/person; 1hr 30min from Miami by car.*

Monarch Butterfly Migration, Michoacán
Come late October, millions of monarch butterflies flutter down from the northern US to blanket the oyamel firs of the Monarch Butterfly Biosphere Reserve, in Mexico's Michoacán state. Watching them mass is a true bucket-list item. *2hr from Mexico City by car.*

Sea Mammals, British Columbia
Cruise the Haro Strait and the Strait of Juan de Fuca from Victoria to downtown Vancouver with your binoculars out. Spot elegant orcas, breaching porpoises and playful seals and sea lions. *www.princeofwhales.com; $215/person; 3hr from Vancouver by car.*

N
0
0

150 km
80 miles

120 miles (195km)

Allentown

NEW
JERSEY

Johnstown

PENNSYLVANIA

08

Reading

Hershey
Harrisburg

07

80 miles (130km)

Lancaster

Trenton
Norristown

York

Philadelphia

10

06

Hagerstown

Wilmington
Newark

MARYLAND

Millville

Frederick

40 miles (65km)

01

WEST
VIRGINIA

04

Baltimore

Dover

Winchester

02

Wildwoods

Leesburg

Rockville

03

WASHINGTON, DC ✪

Annapolis

DELAWARE

Alexandria

VIRGINIA

St Charles

Harrisonburg

05

11

Ocean City

Fredericksburg

12

Charlottesville

09

Richmond

Lynchburg

ATLANTIC OCEAN

Petersburg

The US capital's central location makes for easy and varied getaways. Head south to Virginia for syrupy-paced country life, west to the Appalachian mountains, north to bustling Philadelphia, or east to the Maryland shore for seaside escapes.

WASHINGTON, DC

● ARTS & CULTURE   ● HISTORY   ● OUTDOORS   ● FOOD & DRINK   ● FESTIVALS & EVENTS   ● MUSIC & FILM

——— ONE HOUR FROM ———

01 Baltimore

This harbour city is rapidly leaving behind its reputation for urban decay, with new arts districts, gallery spaces and quirky hotels and restaurants. The old Bromo Seltzer Tower, with its landmark clock face, has been transformed into artists' studios, open Saturdays. Area 405, a 170-year-old former brewery, is another hotspot for emerging artists. Crash at Hotel Revival, with rooms filled with local masterpieces. *1hr by car or bus, 30min by train.*

02 Annapolis

On the Chesapeake Bay and home to the US Naval Academy, Maryland's capital is all about the water. Take a wooden schooner cruise around the bay, or learn to sail yourself at the Annapolis Sailing School. In summer, cheer on the Wednesday night sailboat races launching from Spa Creek. Then gawk at the mega-yachts on 'Ego Alley' while you sip a margarita. *40min by car or 50min by train & bus from Union Station.*

03 Virginia Horse Country

Just an hour outside DC is the impossibly genteel town of Middleburg, long a favoured retreat of Washington elites. The 'nation's horse and hunt capital' is a land of stately country homes and even more stately barns, the latter home to million-dollar racehorses. Come in April for the Middleburg Spring Races or in summer for weekend polo matches. Or anytime for the wineries that dot the nearby hills. *www.middleburg springraces.com; Apr; 1hr by car.*

———TWO HOURS FROM ———

04 Harpers Ferry, West Virginia

At the confluence of the Shenandoah and Potomac Rivers, the green hills and quaint restored churches of Harpers Ferry belie the town's bloody history. Stroll the Civil War battlefield of Bolivar Heights and visit the fort where abolitionist John Brown staged his ill-fated raid. Lighten up with a stroll along the C&O towpath, then rest up at one of the area's delightful historic B&Bs. *1hr 10min by car or 1hr 40min by train from Union Station.*

05 Shenandoah National Park

Less than 90 minutes after escaping the snarl of DC traffic you'll find yourself coasting along the backbone of the Blue Ridge Mountains on Skyline Drive, one of the most glorious roads in America. Running down the centre of Shenandoah National Park, it passes meadows of grazing deer, ancient hardwood forests and the craggy peaks of Hawksbill and Old Rag mountains. Hike, picnic, animal-watch, camp, repeat. *1hr 20min by car.*

06 Berkeley Springs, West Virginia

The mineral waters in these mountains drew Native Americans for thousands of years before they attracted colonials like George Washington. Today, it's a quirky spa town, with a handful of hot spring hotels, galleries and mom-and-pop cafes. Soak for cheap in the historic Roman Bathhouse at Berkeley Springs State Park, a welcome treat after the long hike up Cacapon Mountain, or splash out on a private whirlpool at Renaissance Spa. *1hr 50min by car.*

CIVIL WAR SITES

As a number of the most momentous Civil War sites are an easy drive from DC, it's possible to string several together for a multi-day history tour. First head to Manassas (40min), where, in July of 1861, elite Washingtonians arrived with picnics to watch a glorious Union victory. Instead, they saw a bloody Confederate win. A bit further south is the Fredericksburg and Spotsylvania National Military Park (www.nps.gov/frsp/index.htm; 1hr), commemorating four major nearby battles. Turn north towards Maryland to Antietam (2hr20min), site of the bloodiest one-day battle in US history. Just northeast is Pennsylvania's infamous Gettysburg (1hr).

———— THREE HOURS FROM ————

07 **Amish Country, Pennsylvania**
Buggies still ply the roads of southeastern Pennsylvania, home to the simple-living, modern technology-shunning Amish. So if you're looking to escape the rat race for a few days, this is the place. Shop Lancaster Central Market for Amish-grown veggies and handmade quilts. Chow on traditional treats like apple dumplings and shoofly pie. Ride a buggy or take a farm tour. Then ditch your phone for a farmstay at a real Amish home. *2hr 10min by car.*

08 **Hershey, Pennsylvania**
If you've got kids (or just a sweet tooth), go directly to rural Pennsylvania's Hershey's Chocolate World for a tram ride through a chocolate factory of singing animatronic candies. Afterwards hit the airplane hangar-sized gift shop for Hershey's Kisses the size of your head, then head across the street to the century-old Hersheypark for all-American try-to-make-you-hurl amusement park rides and a seasonal waterpark. *www.hersheypark.com; 2hr 20min by car.*

09 **Monticello**
Thomas Jefferson started building his plantation when he was just 26, and the estate is a testament to his polymath genius. Visit the Palladian-style mansion with its octagonal dome and cannonball-weighted Great Clock, then stroll the lush gardens. Don't miss a visit to the outbuildings once occupied by the enslaved people who worked the land, including Sally Hemings, mother of six of Jefferson's children. *https://home.monticello.org; 2hr 20min by car.*

10 **Philadelphia**
Jump from America's current to former capital for a dose of Revolutionary history. The easily walkable Historic District features 23 sites, including Independence Hall, where the Constitution was created, the famously cracked Liberty Bell, and Christ Church Burial Ground: final resting place of Benjamin Franklin. On fine days, repair to the Delaware River waterfront for drinks, nibbles and boat-spotting. *2hr 30min by car or 2hr by train from Union Station.*

Courtesy of Hotel Revival

01

11 **Ocean City, Maryland**
Sometimes all you want from a summer weekend is a lump of fried dough, a walk on the boardwalk and an airbrushed T-shirt with your own face on it. So go 'down the ocean', as the Marylanders say, to this beloved Eastern Shore beach. Quintessential Ocean City pastimes include fishing, minigolf, all-you-can-eat crab feasts, drinking at tiki bars and riding rickety carnival rollercoasters. Do it all!
2hr 40min by car.

12 **Assateague Island National Seashore**
Spanning the Eastern Shore of Maryland and Virginia, this wild stretch of sea, salt mash and maritime forest is most famous for its roving feral horses. Legend has it these 'Chincoteague ponies' are descended from survivors of a 16th-century Spanish shipwreck, and though the reality is probably more prosaic, the horses are anything but. Watch them running wild from your kayak, bike, campsite or clam-digging spot. *3hr by car.*

© Moelyn Photos / Getty Images

N
0 ——— 150 km
0 ——— 80 miles

15
18
Hudson

120 miles (195km)

MASSACHUSETTS
Worcester
Chicopee

16
Kingston
12
80 miles (130km)

CONNECTICUT
Hartford

RHODE
ISLAND

NEW YORK
11 Poughkeepsie
Newburgh
Waterbury

Norwich

17
New London

09
Port Jervis
Danbury
New Haven

08

PENNSYLVANIA
07
40 miles (65km)
04
Bridgeport

Scranton

West
Orange
Stamford

06
05
14
Sag Harbor

Morristown
03 01
New York City
10
02

Allentown
New Brunswick
13

Reading
Asbury Park

Binghamton

Philadelphia

Wilmington

MARYLAND
Millville

Dover
Atlantic City

DELAWARE

ATLANTIC
OCEAN

New Yorkers know their city is the centre of the universe, but the savviest also know that it is the gateway to a constellation of escapes. In just an hour or three, Gotham is replaced by beaches, mountains, historic sites, pretty villages and more.

NEW YORK CITY

● ARTS & CULTURE   ● HISTORY   ● OUTDOORS   ● FOOD & DRINK   ● FESTIVALS & EVENTS   ● MUSIC & FILM

——— ONE HOUR FROM ———

01 Thomas Edison National Historic Park, West Orange, New Jersey

You can still sense Thomas Edison's brilliance and tireless pursuit of knowledge at his sprawling workshops in West Orange, New Jersey. Visit the actual labs where he and his researchers perfected the light bulb by methodically testing 6000 substances and 3000 designs, beginning in the 1870s. *www.nps.gov/edis/index.htm; 1hr by train & bus from Penn Station.*

02 Rockaway Beach, New York

Surfboards on the subway? A surprising sight, no doubt, but it's a quick subway (or scenic ferry ride) from Manhattan to the break at the white-sand beach off 90th St in the Rockaways. A tight-knit group of Atlantic wave worshippers have revitalised this section of the beachfront. If you've forgotten your board, watch the action from a beachside cafe. *1hr by ferry or 1hr 20min by subway.*

03 Morristown National Historical Park, New Jersey

Success in fighting the American Revolution for the Continental Army was still in doubt during the two winters when troops made camp here, in 1777 and 1779. This national park documents how the thousands of soldiers struggled to survive the war and the elements. Four distinct areas feature original buildings, including General George Washington's headquarters during the long frozen months. *www.nps.gov/morr; 1hr 10min by train from Penn Station.*

04 Westport, Connecticut

Purists prize Connecticut-style lobster rolls, which eschew the mayonnaise and other non-essentials of the Maine version and concentrate on the succulent shellfish meat. You'll find stands selling them near the train station in walkable, nautical Westport. Follow the river and look for the house where F. Scott and Zelda Fitzgerald threw their legendary parties, then pause at one of the town's beaches. *1hr 10min by train from Grand Central Terminal.*

05 North Shore Vineyards, Long Island, New York

The North Fork of Long Island, the spit of land that arrows out into Long Island Sound, is known for its bucolic farmland and vineyards. Two very drinkable examples: Pugliese Vineyards, which has been producing sparkling wine since 1980; and Lenz Winery, which creates European-style wines. Pretty Route 25 is edged with farm stands and vineyards; the less-travelled Route 48 also has many wineries. *www.pugliesevineyards.com; www.lenzwine.com; 1hr 30min by car.*

06 Sagamore Hill National Historic Site, Oyster Bay, New York

This sprawling Victorian home is where President Theodore Roosevelt and his wife raised six children and entertained luminaries from around the world. You can sense Roosevelt's irrepressible personality in the decor and possessions on display in every room. A nature trail starting behind the museum ends at a picturesque beach. In summer, reserve house tours online. *www.nps.gov/sahi; 1hr 30min by train & taxi from Penn Station.*

THE
LOCAL'S
VIEW

'I like to escape without having to rent a car: by train, bus, ferry, plane, anything to avoid standing in line to pick up a car, followed by sitting in a line of traffic to get off Manhattan. Hauling luggage by subway forces you to pack a little lighter, leaving as jetsam 'might-as-wells' that would have otherwise made the final cut for a car trip. Just a couple of examples of places easily reached from Manhattan: 17th-century Kingston in the Hudson Valley by express bus, and the iconic beaches of Fire Island by train and ferry.'

Maura Murphy, attorney

07 **New Jersey State Fair, Augusta, New Jersey**

New Jersey isn't all industry, Newark, the Shore or endless suburbs, it's also huge tracts of rural enclaves where farming and horse-rearing still define local life. Each summer, these sylvan pursuits are celebrated at the state fair in pastoral Sussex County. Show horses compete for ribbons, kids show off their prize rabbits and top produce is celebrated, plus there are rides and concerts. ***www.njstatefair.org; early Aug; 1hr 30min by car.***

──── TWO HOURS FROM ────

08 **Pizza, New Haven, Connecticut**

New Haven is home to a unique and celebrated style of pizza (as well as a certain Ivy League university). Thin and cracker-crisp pies are cooked in coal-burning ovens at famed purveyors, some of which date back decades. Although you have your choice of toppings, white clam is the one to go for. Legendary restaurants include Frank Pepe (since 1925), Modern Apizza (1934), Zuppardi's Apizza (1934) and Sally's Apizza (1938). ***1hr 40min by train from Penn Station.***

09 **High Point State Park, Sussex, New Jersey**

The wonderful panoramas at this park include the Poconos to the west, the Catskills to the north and the Wallkill River Valley to the southeast. A towering monument marks the highest point in the park (and in New Jersey) at 1803ft. Trails in the park snake off into the forests and there's a small beach with a lake to cool off in during the summer. ***www.state.nj.us/ dep/parksandforests/parks/highpoint; 1hr 40min by car.***

10 **Jones Beach, New York**

Jones Beach is 6.5 miles of glittering, clean sand covered with bodies. Its character differs depending on which 'field' you choose – for example, 2 is for the surfers and 6 is for families – but it's a vibrant scene no matter where you spread your blanket. Although the beach is the obvious focus, don't miss the acclaimed architecture and design of master-builder Robert Moses. ***www.parks.ny.gov/parks/ jonesbeach; 2hr by train and bus from Penn Station.***

11 **Poughkeepsie, New York**

Near Poughkeepsie train station is the entrance to a former railroad bridge crossing the Hudson River. It's now the world's longest pedestrian bridge – 1.28 miles – and a state park. The soaring span provides breathtaking views up and down the river valley. Don't miss the 3.6-mile loop walking trail across this bridge and back along the Mid-Hudson Bridge to Poughkeepsie. *www. walkway.org; 2hr by train from Grand Central Terminal.*

12 **Hyde Park, New York**

The town of Hyde Park in the Hudson River Valley is a sort of American *Downton Abbey* district, lying between the lavish estates (now national parks) of two great dynasties. President Franklin D. Roosevelt's vast riverside home is a good place to start. His wife, Eleanor, created her own home, the cosier Val-Kill, a few miles inland. Nearby is the estate of the Vanderbilt moguls. *www.nps.gov; 2hr by train & bus from Grand Central Terminal.*

NEW YORK TRANSIT

New York is the hub of the most extensive transit systems in the US. From Manhattan you can reach hundreds of points on the compass, all without the hassle of a car. Amtrak, the Long Island Railroad, Metro North and NJ Transit all run trains from Penn Station and Grand Central Terminal, while a plethora of bus lines (many express) start at the Port Authority Bus Terminal. Check schedules and routes online. For nearby getaways, look to the subway and ferry lines.

10

© Cultura RM / Alamy Stock Photo

13

 13 Asbury Park, New Jersey

Hard by the Atlantic Ocean, Asbury Park is where New Jersey's troubadour, Bruce Springsteen, got his start in the mid-1970s. The first stop for everyone is the Stone Pony, the bar where he launched his career. Another 40 bars, many with live music, lure trains full of young NY-based revellers. Blocks of restored Victorian homes and newer units are also attracting New Yorkers' attention. *2hr by train from Penn Station.*

14 Sag Harbor, New York

The old whaling town of Sag Harbor, on Long Island's Peconic Bay, is edged with historic homes, and its main street is lined with restaurants and shops. Wander along the pier and have a lobster roll. Stop into the Sag Harbor Whaling & Historical Museum to ponder actual artefacts from 19th-century whaling ships. Nearby parks have walking trails through dunes and forests. *www.sagharborchamber.com; 2hr 30min by express bus.*

15 Grey Fox Bluegrass Festival, Oak Hill, New York

For four days in the peak of summer, all manner of uniquely American music, from bluegrass to Cajun, is heard in the little village of Oak Hill in the lush Catskills Mountains. Thousands of fans camp on the grassy fields and jam to the likes of the Del McCoury Band, Peter Rowan and the Steep Canyon Rangers. *www.greyfoxbluegrass.com/festival; late Jul; 2hr 30min by car.*

16 Woodstock, New York

A minor detail: the 1969 music festival was actually held in Bethel, an hour's drive

14

west. Nonetheless, the perfectly quaint town of Woodstock still attracts an arty, music-loving crowd and cultivates the free spirit of that era, with rainbow tie-dye style and local grassroots everything, from radio to a respected indie film festival and a farmers market (fittingly billed as a farm festival). *www.woodstockchamber.com; 2hr 30min by car.*

17 **Sailfest, New London, Connecticut**
During its golden age in the mid-19th century, New London was a large whaling centre and a wealthy port city. Decades of decline followed as whaling waned, but the era is still recalled in the historic downtown. The maritime heritage is celebrated every year at Sailfest, a three-day summer festival featuring live entertainment, tall ships and fireworks over the river. *https://sailfest.org; mid-Jul; 2hr 30min by train from Penn Station.*

──── THREE HOURS FROM ────

18 **Olana, Hudson, New York**
Olana is one of the finest of the Hudson Valley mansions – its owner, acclaimed landscape painter Frederic Church, designed every detail. Inspired by his travels in the Middle East and his love of the views across the Hudson to the Catskills, the 'Persian fantasy' house is extraordinary as is the interior, hung with many of Church's paintings. *www.olana.org; mid-Jun-Oct; 3hr by train from Penn Station.*

© James Kirkikis / Shutterstock

HAMPTONS CHIC

The fabled Hamptons, summer getaway for the elite, comprise a series of villages that perch along eastern Long Island's beaches. In summer, star-spot names from gossip sites, lounge, party and dine at multimillion-dollar estates and exclusive resorts. Many arrive by helicopter from Manhattan, eschewing the commoners on the express train (around 3hr). Summer prices can make the Hamptons off limits for mere mortals, but from fall to spring, rental and hotel rates plummet, and a sandy escape becomes affordable and relatively crowd-free.

© solepsizm / Shutterstock

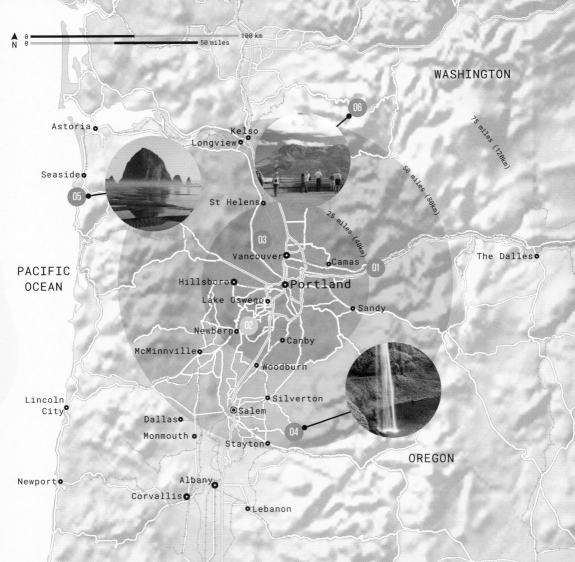

N

0 100 km
0 50 miles

WASHINGTON

06

Astoria

Kelso
Longview

Seaside

05

St Helens

PACIFIC
OCEAN

03

Vancouver

Camas

01

The Dalles

Hillsboro

Portland

Lake Oswego

Sandy

Newberg

02

Canby

McMinnville

Woodburn

Silverton

Lincoln
City

Dallas

Salem

Monmouth

04

Stayton

OREGON

Newport

Albany

Corvallis

Lebanon

50 miles (80km)

25 miles (40km)

75 miles (120km)

For outdoors types longing for beaches, mountains, waterfalls and wine, Portland is bliss: a city in the centre of some of the most beautiful and wild country in the US. So, fuel up on food-cart fare and cups of Stumptown joe, and strap on your boots.

PORTLAND

ARTS & CULTURE   HISTORY   OUTDOORS   FOOD & DRINK   FESTIVALS & EVENTS   MUSIC & FILM

——— ONE HOUR FROM ———

01 Waterfall Alley, Columbia River Gorge

Steep, evergreen-covered cliffs are striated with so many high, lush waterfalls that it boggles the mind. Walk into this temperate rainforest and you'll find golden-red foliage in fall, snow sprinkles in winter, wildflowers in spring and flourishing mosses year-round. The lower edge is dominated by the mighty, curving Columbia River. Adventures range from easy jaunts from a parking lot to full days exploring the wilds. *www.fs.usda.gov/crgnsa; 30min by car.*

02 Willamette Valley Wine Country

The New World's answer to Burgundy, the bucolic Willamette Valley excels at producing complex pinot noirs. Small towns chock-full of gourmet restaurants and cute shops are separated by rolling hills covered in grapevines and flowers. Most wineries are family-owned and you're more likely to be met by a happy dog than an attitude. Beyond wine, check out distilleries, cheese factories, bike trails and more. *40min by car.*

03 Sauvie Island

With pick-your-own fruit farms in summer, corn mazes and hay rides come autumn, and beaches, trails and waterways to explore year-round, Sauvie Island is an easy trip to pastoral bliss. Plus, it's as fun for kids as it is for adults. As the largest island in Columbia River, any land not dedicated to farming is a protected wildlife area. Expect to see ospreys, herons, bald eagles and maybe even a beaver. *40min by car.*

04 Silver Falls State Park

Quite possibly the most stunning park in the state, the 7.2-mile, not-to-miss Trail of Ten Falls does what it says and more. Walk the full loop to see all the waterfalls in their fern-draped, forested splendour, or park at the Silver Falls Lodge from where it's a short walk down to South Falls – here the trail leads underneath an unforgettable 177ft cascade curtain. *www.oregonstateparks.org; 1hr 30min by car.*

05 Cannon Beach

The Oregon Coast holds some of the most beautifully grand but lesser-known beaches in the contiguous USA. Cannon Beach is the ideal, easy-to-reach introduction to the region, with miles of white sand, waterfalls tumbling to the beach and the hulking, 72m Haystack Rock which makes for a picture-perfect silhouette. In town, you'll find art galleries, chic boutiques and incredible dining (hint: try the chowder). *1hr 30min by car.*

——— THREE HOURS FROM ———

06 Johnston Ridge Observatory, Mt St Helens

Gaze into the awesome crater of Mt St Helens from this state-of-the-art observatory. Whether this is your destination or you're preparing for a longer hike through the National Volcanic Monument, you'll learn the science behind the 1500 atomic bomb-force eruption in 1980. Then take the 1-mile Eruption Trail walking loop for even more views, with explanations, of the volcanic destruction. *www.fs.usda.gov; closed winter; 2hr 30min by car.*

TIMBERLINE LODGE

Stay, dine, hike, ski or gawk – this iconic Oregon lodge (*www. timberlinelodge. com; 1hr 30min by car*) is a destination in itself or a must-stop if you're exploring Mt Hood. The grand structure was built of local stone and timber to blend in with the surrounding forest and the six-sided central tower is meant to mimic the pyramid-like peak of its glorious mountain location. It's a National Historic Landmark as well as a hotel, a fine restaurant, a pub, a ski resort and a trailhead for adventurous hiking. You may recognise the exterior from the 1980s horror film, *The Shining*.

BEST FOOD & DRINK TOUR

Food plus a road trip is as classic an American combo as peanut butter and jelly.

Est. 1888

FourRoses
DISTILLERY

1224 Bonds Mill Road

S IN NORTH AMERICA

Bourbon Trail, Kentucky

Grab a designated driver and wind through the bluegrass hills, stopping at the dozen-plus distilleries that dot the state. Tour the facilities and sample the golden liquor, made from corn and aged in oak barrels here since the 1800s. *www.kybourbontrail. com; 1hr from Louisville by car.*

BBQ Tour, Central Texas

The air of Central Texas is perfumed with smoky meat; if you're not sold on BBQ here, vegetarianism might be for you. Gorge on succulent pork ribs at Luling's City Market or weep over the perfect brisket at Taylor's Louie Mueller. *1hr from Austin by car.*

Clam- & Oyster-Digging, Willapa Bay, Washington

Wade into the gooey mudflats of Washington's Willapa Bay to dig for razor clams, cockles, littlenecks and sweet, fat oysters. Then grill them up or just suck them down. *3hr from Portland, Oregon by car.*

Hatch Chile Festival, New Mexico

Come late summer, New Mexico is fragrant with the scent of roasting chiles, the region's most beloved foodstuff. Taste them at this September festival, with music, chile wreath-making workshops, a chile-eating contest and more. *www.hatchchilefest. com; Sep; $20 per car; 3hr from Albuquerque by car.*

Maple Open House Weekend, Vermont

Every March, the 'sugar shacks' of Vermont open their doors in honour of the year's first maple syrup harvest. Gorge on maple creemees (maple soft serve ice cream), maple donuts, maple-boiled hot dogs, and the unbeatable toffee-like maple-on-snow. *www.vermontmaple. org/maple-open-house-weekend; 2hr from Boston by car.*

State Fair, Wisconsin

You better fast for all of July, 'cos come August this epic state fair means face-sized cream puffs (a speciality since 1924), cherry pie on a stick, maple cotton candy and the iconic deep-fried cheese curds. *www.wistatefair.com/ fair; Aug; 1hr from Madison by car.*

Peach-Picking, Georgia

Nothing says Southern summer like a day spent out in the orchard, picking sugar-sweet peaches from a sun-warmed tree. Try Southern Belle Farm, with a country market and friendly resident donkeys. *www.southernbelle-farm.com; 1hr from Atlanta by car.*

Campbell River Salmon Fishing, British Columbia

Known as the 'salmon capital of the world', this postcard-worthy British Columbia town is a prime spot for catching (and eating) all five salmon species – chinook, coho, sockeye, pink and chum – at different times of year. *3hr from Victoria by car.*

Cajun Country Boudin Trail, Louisiana

New Orleans gets all the glory, but it's out in rural Cajun Country where food gets really wild. Follow the 'boudin trail' to authentic eats: boudin (a pork and rice sausage), hog's head cheese, and fried meat pies. *www.cajunboudintrail. com; 2hr from New Orleans by car.*

Mole-Making, Puebla

This colonial jewel box of a city is the spiritual home of mole, the famously complex Mexican sauce of nuts, fruit, chocolate and spices. Try your hand with a class at Mesón Sacristía, a boutique hotel and cookery school. *www.mesones-sac-ristia.com; 2hr from Mexico City by car.*

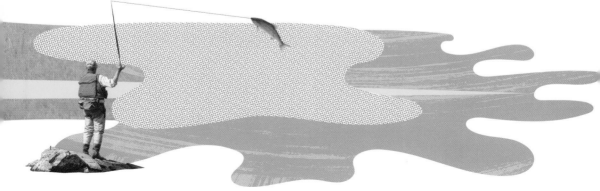

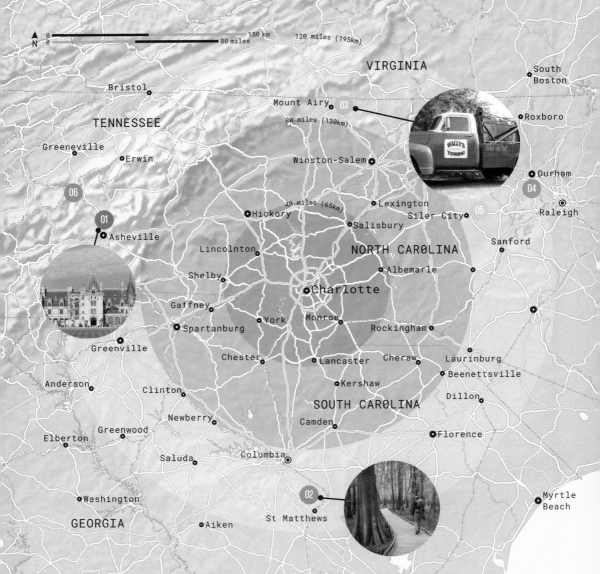

CHARLOTTE

Beyond the South's financial capital, discover the laid-back hospitality of the Carolinas. From arty mountain towns to small-city film festivals, and chilled-out hot springs to frisky fireflies, getaways from Charlotte are quirky Americana at its best.

VIRGINIA

TENNESSEE

NORTH CAROLINA

SOUTH CAROLINA

GEORGIA

South Boston
Bristol
Mount Airy 03
80 miles (130km)
Roxboro
Greeneville
Erwin
Winston-Salem
Durham
06
40 miles (65km)
Lexington
04
01
Hickory
Salisbury
Siler City
Raleigh
Asheville
05
Lincolnton
NORTH CAROLINA
Sanford
Shelby
Albemarle
Gaffney
Charlotte
York
Monroe
Spartanburg
Rockingham
Greenville
Chester
Lancaster
Cheraw
Laurinburg
Anderson
Kershaw
Beenettsville
Clinton
Dillon
Newberry
SOUTH CAROLINA
Elberton
Greenwood
Camden
Florence
Saluda
Columbia
Washington
Myrtle Beach
GEORGIA
02
Aiken
St Matthews

N
0
0
150 km
80 miles
120 miles (195km)

● ARTS & CULTURE   ● HISTORY   ● OUTDOORS   ● FOOD & DRINK   ● FESTIVALS & EVENTS   ● MUSIC & FILM

—— TWO HOURS FROM ——

01 Asheville, North Carolina
In 1890, George Vanderbilt decided to build himself the largest home in the US in the most beautiful location. He chose the rolling mountains near Asheville – along what's now the Blue Ridge Parkway – where art deco buildings, farm-to-table restaurants and innovative microbreweries now pave the way to his Biltmore Estate. The town is home to a sizable population of artists and hard-core hippies. **www.biltmore.com; 2hr by car.**

02 Congaree National Park, South Carolina
Firefly speed dating? Yup. Not only does Congaree National Park have the world's largest expanse of old-growth hardwood forest and miles of boardwalk trails through languid blackwater swamps, but for two weeks in late May to early June, Congaree becomes a first-date hotspot for a million Photinus carolinus fireflies, who blink in unison during mating season. **www.nps.gov/cong; 2hr by car.**

03 Mount Airy, North Carolina
The inspirational setting for the *The Andy Griffith Show*'s fictional town of Mayberry, Mount Airy is as American as the $1.80 bologna sandwich at its Snappy Lunch diner. The surrounding Yadkin Valley's wine industry has lent an air of sophistication to this old-timey town. Check into the endearingly perfect Pilot Knob Inn to stay in an original tobacco cabin, and canoe on your own private lake. **www.pilotknobinn.com; 2hr by car.**

04 Full Frame Documentary Film Festival, Durham, North Carolina
In early April, smart Durham opens its historic downtown theatre, tobacco warehouse entertainment district and down-home barbecue joints to thousands of documentary film-makers and film-goers. The eclectic, Southern-liberal Bull City turns into a city-wide screening party, with over 100 films, talks and events. **www.fullframefest.org; early Apr; 2hr by car or 2hr 30min by train.**

05 Celebrity Dairy, Chatham County, North Carolina
Meet new human friends over one of Celebrity Dairy's goat's cheese-fuelled, seven-course monthly suppers, or visit goat kids during Open Barn weekends (usually February, March and November). Stay overnight in the farmhouse (which surrounds a 200-year-old log cabin) to allow time for window-shopping in historic Pittsboro, or catch a band in nearby college-town Chapel Hill. **www.celebritydairy.com; 2hr by car.**

06 Hot Springs Resort & Spa, North Carolina
After hiking two city blocks of the Appalachian Trail that runs right through Hot Springs, relax in a mineral-spring-fed hot tub on the French Broad River, roast s'mores over a campfire, or get adventurous in a white-water raft. Lodging options include primitive campsites on the river, rustic cabins, fully outfitted rooms and luxury suites, complete with heart-shaped tubs where you can bathe privately in the healing waters of Hot Springs. **www.nchotsprings.com; 2hr 30min by car.**

THE LOCAL'S VIEW

'Greensboro is great for history. Visit the International Civil Rights Center & Museum located in an old Woolworth's, with the famous lunch counter. Settled by Quakers, Greensboro was a terminus of the Underground Railroad (a network of secret routes set up in the 19th-century for African-American slaves to escape into free states). In the woods on Guilford College's campus, a trail simulates what slaves went through. Leave space in your vehicle for a treasure from Architectural Salvage. You may find a 100-year-old chandelier to take home.'

Rose Hoban, reporter & founder of North Carolina Health News

The Bay Area and beyond offers a stunning array of sights and activities, from wildlife-viewing on the seashore and wine tasting among vineyards to exploring old ghost towns in the Gold country and walks through groves of redwood trees.

SAN FRANCISCO

ARTS & CULTURE   HISTORY   OUTDOORS   FOOD & DRINK   FESTIVALS & EVENTS   MUSIC & FILM

——— ONE HOUR FROM ———

01 Sausalito

Just across the Golden Gate Bridge, Sausalito's picturesque town centre has a charming strip of art galleries, ice cream parlours, cafes and boutiques, most set inside pretty pastel-painted Victorian buildings. Just beyond the town is a fascinating artistic houseboat community. Fort Baker, once a military garrison, is another worthwhile stop for a visit to the kid-friendly Bay Area Discovery Museum. ***30min by car or ferry.***

02 Muir Woods

Forget San Francisco's skyscrapers, head north to marvel at Mother Nature's towering giants, a grove of trees that soar over 250ft. Muir Woods, named after the naturalist John Muir and established as a national monument in 1908, is a stunning slice of nature where you can walk between the massive redwoods and commune with an old-growth forest. Advance parking and shuttle reservations are required. ***www.nps.gov; 30min by car.***

03 Half Moon Bay

As one of three self-described 'Pumpkin Capitals' in America, Half Moon Bay is a great place to visit in October. The rest of the year, visitors stop here to horseback ride or cycle along the coast, or (in winter) to check out the famed surf spot Mavericks. At Pillar Point Harbor you can see crabs being sold straight from the fishing boats, and enjoy fried calamari at Barbara's Fish Trap. ***30min by car.***

 04 Berkeley

A national hotspot of (mostly left-of-centre) intellectual discourse and birthplace of the free-speech and counterculture movements, Berkeley still has a hippie vibe alive on Telegraph Ave, where you can shop for tie-dye T-shirts, browse vinyl at the Rasputin record store and listen to Grateful Dead songs played by street buskers. Don't miss a wander around the UC Berkeley campus and a visit to the remodelled Berkeley Art Museum and Pacific Film Archive. ***40min by car or BART.***

05 Palo Alto

A green oasis in the heart of Silicon Valley, Palo Alto is worth a visit for a peek inside the tech industry. Start with a walk past the HP garage, where Hewlett-Packard had its modest start, then stroll down University Ave where venture capitalists lunch with internet entrepreneurs. More conventional sights include the quirky Junior Museum & Zoo, the beautiful gardens at the Elizabeth F Gamble home, and the Cantor Center for Visual Arts at Stanford University. ***40min by car.***

06 Sonoma Valley

It was in Sonoma in 1846 that American rebels revolted against the Mexican government, an incident known as the 'Bear Flag Revolt'. Wander around the historic town square, the largest plaza in California, home to excellent restaurants, shops and the Mission San Francisco Solano. Kids will love the Sonoma TrainTown Railroad with its miniature locomotives. Later, head north on Hwy 12 to visit wineries and Jack London Historic State Park. ***1hr by car.***

SF TRANSPORT

Surrounded by water on three sides, San Francisco's transport options for escape include bridge, ferry or train. If you are heading north to wine country or the Mendocino coast, cross over the Golden Gate Bridge by car or bus and travel onwards along Hwy 101. You can also take the ferry from downtown SF to Sausalito or Larkspur for bus and train connections. For Palo Alto and the South Bay, take Caltrain from the 4th and King station; for East Bay locations such as Oakland and Berkeley, use BART, the Bay Area's subway system. You can also reach Oakland by ferry. For most escapes, though, your best bet is to rent a car.

© Mark Read

07 The Barlow, Sebastopol

Four blocks of restaurants, cafes, craft breweries and markets make The Barlow district an extravaganza for foodies wanting to taste locally produced wine, cheese and other delicacies. Much of what's available here is farm-to-table, and some ingredients are available in the community supermarket. The town of Sebastopol, home to artists, writers and New Age folks, is a fine place to explore boutiques and galleries. *www.thebarlow.net; 1hr by car.*

08 Duarte's Tavern, Pescadero

Duarte's Tavern has been a fixture in Pescadero since 1884, serving hearty American fare with Portuguese influences. Most of the vegetables come from the garden out back, including ingredients for its justifiably famous artichoke soup. After filling up, explore the village, the old barns, pastures and nearby coast. Just down the road is Pigeon Point Lighthouse, where you can overnight in a cabin, and Ano Nuevo, home to thousands of barking elephant seals. *www.duartestavern.com; 1hr 10min by car.*

09 Charles Schulz Museum, Santa Rosa

Fans of the Peanuts comic strip can learn all about Charlie Brown, Snoopy and the gang at this museum, through videos, exhibits and a replica of Charles Schulz's office. Next door you'll find more family fun at the Santa Rosa Children's Museum. Also nearby is the quaint town of Petaluma, which has served as the backdrop for films including *American Graffiti* and *Basic Instinct*, plus Ronald Reagan's 1984 *Morning in America* commercials. *www.schulzmuseum.org; 1hr 10min by car.*

10 Gilroy Garlic Festival

During the last weekend in July, garlic-lovers flock to the otherwise sedate town of Gilroy for a raucous three days of eating and indulging in everything garlic. The food available is legendary and pungent – don't miss the garlic ice cream. The highlight here is Gourmet Alley, where daredevil 'pyro-chefs' mix large skillets of garlic-infused shrimp and scampi. Bring an empty stomach and mints! *www.gilroygarlicfestival.com; Jul; 1hr 20min by car.*

11 BottleRock Music Festival, Napa Valley

One of Northern California's premier music festivals, the four-day long BottleRock has been attracting some of the world's biggest rock bands, including The Killers and the Red Hot Chili Peppers, since the inaugural event in 2013. After a few days of rocking out, you can switch gear and explore the upscale wineries of Napa Valley. Consider taking the wine train, which travels from Napa to St Helena, stopping at the world's most famous wine producers. *www.bottlerocknapavalley.com; May; 1hr 20min by car.*

The Russian River

Just an hour north of San Francisco you can float down the gentle Russian River on kayaks or inner tubes, overnight in rustic cabins or simply lounge on the beach with a six-pack of Russian River Brewing Company beer. After sunbathing on Johnson's Beach in Guerneville, head into the nearby town of Healdsburg and make for its shady town square, considered one of the most beautiful in America. *1hr 30min by car.*

San Juan Bautista

Founded as a Spanish mission in 1797, San Juan Bautista was home to around 1200 native converts in its day and still serves the local community. There's lots to see around the mission and historic park, including period buggies, stables, an antique hotel and an old saloon. Next up: an epic day of hiking and rock climbing amid the stone outcrops of Pinnacles National Park, 35 miles south. *www. oldmissionsjb.org; 1hr 30min by car.*

GOURMET'S DELIGHT

The hamlet of Yountville (*1hr 20min by car*) has a population of just 3000 but an impressive two Michelin-starred restaurants, the French Laundry and Bouchon, both opened by celebrated chef Thomas Keller. Be prepared for a memorable (and pricey) meal at the more famous French Laundry, where the menu is constantly changing but focuses on gourmet French cuisine. A third Keller restaurant, Ad Hoc, serves comfort food at relatively more affordable prices. Stay overnight at one of Yountville's cosy B&Bs and go for a morning cycle along the Napa Valley bike trail or see the valley from 1000ft from a hot air balloon.

© Kevin Thrash / Getty Images

Santa Cruz

For some good old American fun, nothing beats the Santa Cruz Beach Boardwalk and its classic wooden rollercoaster, saltwater taffy and carnival games. Surfers will find joy at Steamer Lane and the nearby Santa Cruz Surfing Museum, the first of its kind. Kids will love a visit to the Roaring Camp, where historic steam trains trundle through a redwood forest. *1hr 30min by car.*

© Kris Davidson

14

────── TWO HOURS FROM ──────

Monterey Bay

The world-class Monterey Bay Aquarium demands a full day to explore its shark tanks, tide pools and other marine exhibits. But Monterey is worth a long weekend to spend time relaxing in the village of Carmel-by-the-Sea, chock-full of art galleries, cafes and clothing boutiques. The private coastal road connecting Carmel and Monterey is 17-Mile Drive, with stunning views of sandy beaches, wildlife and coastal cliffs. *2hr 30min by car.*

────── THREE HOURS FROM ──────

Columbia

One of the best-preserved gold-rush-era towns in California, Columbia, founded in 1850, was in its heyday the state's second largest city. Today it's a perfectly preserved mining town, where you can pan for gold, ride on a horse-drawn carriage and dress up in 1850s garb for family photos. From Columbia, you can travel north along historic Hwy 49 to visit more Gold Country towns, including Murphys and Placerville. *3hr by car.*

© Katie Jordan / Getty Images

15

Mendocino

Trapped in a time warp, this village of whitewashed houses on a dramatic bluff overlooking the Pacific Ocean has developed into a popular artists' colony. Mendocino is home to several galleries, walking trails and beaches and has been the backdrop of numerous movies and TV shows, including *Karate Kid* and *Murder, She Wrote*. In nearby Fort Bragg you can board the Skunk Train for an historic ride into a redwood forest. *3hr 20min by car.*

 Lake Tahoe

Of the dozen or so ski resorts around Lake Tahoe offering some of the best skiing in the country, the best-known is Squaw Valley, which hosted the 1960 Winter Olympics. There are still many opportunities for adventure for summertime visitors: jet-skiing or waterskiing on Lake Tahoe, white-water rafting on the Truckee River or careering down the bare ski slopes on a mountain bike. *3hr 50min by car.*

SPACE & TECHNOLOGY

The NASA Ames Exploration Center in Santa Clara (*www.nasa. gov/ames; 50min by car*) offers an excellent peek into America's history of space exploration. Exhibits include 1960s-era space capsules, a moon rock, a space station mock-up and film footage of NASA's exploration of Mars and other planets. The hangars near the museum are some of the world's largest freestanding structures. Just 13 miles southeast of NASA is the Tech Museum of Innovation (*www.thetech. org; 1hr by car*) in downtown San Jose, where you can learn more about space exploration, as well as robotics, medical science and other technological wonders.

© Matt Munro

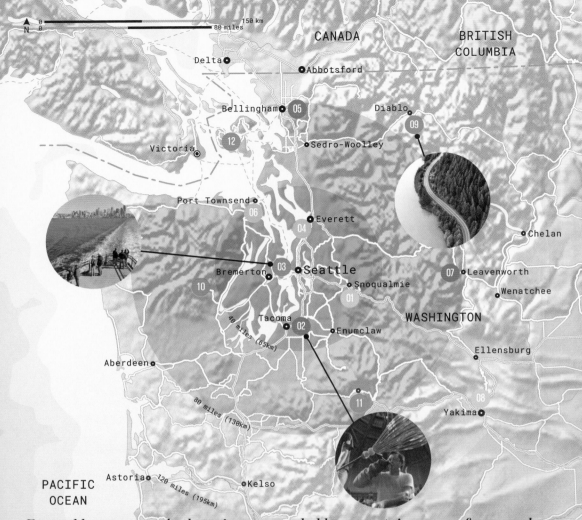

CANADA

BRITISH
COLUMBIA

Delta

Abbotsford

Bellingham **05**

Diablo

09

Sedro-Woolley

12

Victoria

Chelan

Port Townsend

06

04 Everett

Leavenworth
07

Bremerton **03** Seattle

Snoqualmie

Wenatchee

10

01

WASHINGTON

Tacoma

02 Enumclaw

Ellensburg

40 miles (65km)

Aberdeen

80 miles (130km)

11

Yakima **08**

Astoria 120 miles (195km)

Kelso

PACIFIC
OCEAN

Fronted by ocean and otherwise surrounded by mountain ranges, farms and a lauded wine region, Seattle is a base for many delights. Hike national parks, dine farm-to-table, paddle with whales, shop arty boutiques and peruse odd museums.

SEATTLE

● ARTS & CULTURE   ● HISTORY   ● OUTDOORS   ● FOOD & DRINK   ● FESTIVALS & EVENTS   ● MUSIC & FILM

—— ONE HOUR FROM ——

01 Snoqualmie Valley
Seat of farm-to-table dining, dairy farms, lush orchards and small-town Americana, this mountain-framed valley is as lovely a drive as it is a destination. Head to Duvall for boutique shopping and wine tasting, or to North Bend for *Twin Peaks* nostalgia and a slice of cherry pie at Twede's. Don't miss gaping at 82m Snoqualmie Falls from Salish Lodge in Snoqualmie town. *30min by car.*

02 Museum of Glass, Tacoma
Tacoma native Dale Chiluly's fantastical creations are featured heavily at this exceptional museum dedicated to glass art. There are changing exhibitions and some permanent installations, including the water-like Fluent Steps that spans a 64m reflecting pool. Don't miss the Hot Shop where you can watch artists create works from molten glass. The extraordinary Bridge of Glass links to the Museum of Natural History. *www.museumofglass.org; 40min by car.*

03 Bainbridge Island
The ferry trip across the Puget Sound to this laid-back, forested isle immediately gets you into nature. Enjoy the views back to Seattle with Mt Rainier looming in the distance, and breathe in the fresh air. On the island, laze in waterfront cafes, hit the beaches, taste local wines or rent a bike to explore more of the mostly flat countryside. *50min by ferry.*

04 Future of Flight Aviation Center & Boeing Tour, Mukilteo
Learn about the history of aviation and admire old aircraft at the museum before the taking the tour to watch Boeing planes on the assembly line. You get to go into the real workplace (with guides) to watch how the aeroplanes are built. Tours are 90 minutes' long and best and busiest on weekdays when most of the workers are present. *www.futureofflight.org; 40min by car.*

05 Bellingham
One of the most charming coastal towns in the state, Bellingham melds a lively student population with outdoor sports fanatics and liberal thinkers. Hang out at an arty cafe or restaurant, stroll the waterfront in the leafy, Victorian Fairhaven neighbourhood or hook up with a kayak tour. If you're driving, definitely detour along Chuckanut Drive, with oyster bars along the way and views of the San Juan Islands. *1hr 30min by car.*

——TWO HOURS FROM ——

06 Port Townsend
Sitting in rain shadow, this oft-sunny town is a delight. Wander downtown to visit quirky art galleries, fun boutiques and fine restaurants in manicured Victorian buildings, then hit Fort Warden to visit a lighthouse, marine science centre and learn about the area's shipping and military history. You can also join whale-watching excursions or rent kayaks from the town's waterfront. Festivals fill the summer calendar. *2hr by car & ferry.*

07 Leavenworth
Bavaria too far? Thank goodness for Leavenworth, where you can eat German sausages, quaff frosty beers and see enough

SEATTLE FROM THE PUGET SOUND

Hopping on a ferry (*www.wsdot. wa.gov/ferries*) to anywhere is likely to be one of your most memorable escapes from Seattle's concrete jungle. The sound of traffic is replaced by seagull squawks and lapping waves and there's no better view of the city than from on deck, with white-capped Mt Rainier framing the metropolis in the distance. There are lots of destinations to chose from as close as West Seattle (*10min*) to as far away as Victoria BC, Canada (*2hr 50min*). Local fares are cheap enough that you could hop on and do a round trip, just for the ride.

© imageBROKER / Alamy Stock Photo; © franckreporter / Getty Images; © Justin Kuravackal / Museum of Glass

men in lederhosen to save that fare to Europe. The mountain scenery feels like *The Sound of Music* and you'll actually hear Bavarian music piped through the streets. Visit the Nutcracker Museum or take off on myriad outdoor activities in the surrounding area. Christmas here is magical. *2hr 20min by car.*

  Yakima Valley Wine Country
It's dubbed 'The Palm Springs of Washington' for its sunshine but the most extraordinary aspect of this valley is the wine. In between towns, including 1950s-cute Prosser and Wild West–style Native American Toppenish, are hills of Bordeaux grapes, rustic winemaker sheds and giant chateaux, all contributing to some of the best reds in the country. It's an excellent road trip with a designated driver. *2hr 30min by car.*

09 Diablo Lake
This intensely turquoise lake sits in a bowl ringed by mesmerising and often snow-topped peaks. There are heaps of trailheads to view-filled hikes, or plop in a canoe for paddling or fishing for rainbow trout. You can book day trips from Seattle or educational tours with North Cascades Environmental Learning Center. Don't miss the Diablo Dam Overlook for one of the best views in the Cascade Mountains. *www.ncascades.org; 2hr 30min by car.*

10 Olympic National Park
The easy-to-reach Staircase side of Olympic National Park is dominated by towering Douglas fir trees, the icy blue rapids of the Skokomish River, rugged peaks, vibrantly green mosses and so many

hiking trails. If angling or boating is more your thing, tranquil, 4000-acre Lake Cushman is right next door, and there's plenty of camping if you decide to stay longer. *www.nps.gov/olym; 2hr 30min by car.*

——— **THREE HOURS FROM** ———

11 Paradise, Mt Rainier
This glacier-clad volcano is so tall (4392m) that it can be seen from most of the state – but even that can't prepare you for the beauty you'll find here. Flowery meadows, misty views over endless mountain scenery, glacier-filled valleys, rivers, bears... Aptly named Paradise has a visitor's centre, a museum and several trailheads, making it the perfect entry point to this spectacular national park. *www.nps.gov/mora; 3hr by car.*

© Anton Foltin / Shutterstock

12 San Juan Island

Orca sightings, vineyards, forested hills, farmlands and a distinct maritime-bohemian vibe await on this jewel-like island. On land, there's an 1860s military camp, hiking trails, the adorable town of Friday Harbor, wineries and top-notch restaurants to enjoy. Or get on a kayak or a whale-watching tour to get close to the abundant marine life. You may even spot orcas from the ferry. *3hr 30min by car & ferry.*

© roclwyr / Getty Images

© Richard A McMillin / Shutterstock

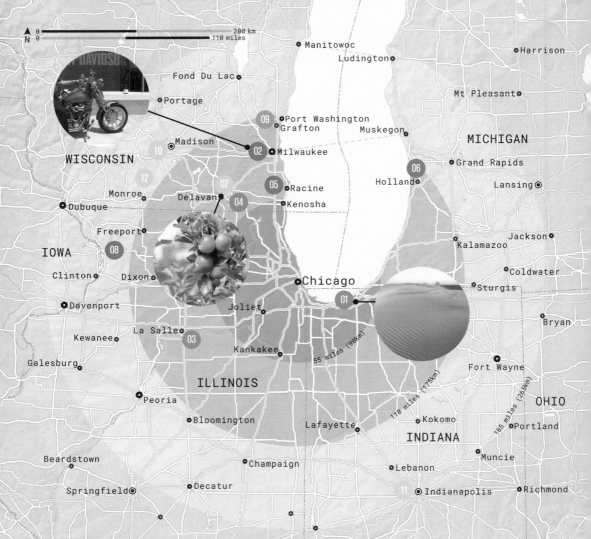

N

0 ——————— 200 km
0 ——————— 110 miles

Manitowoc
Ludington
Harrison

Fond Du Lac
Portage
Mt Pleasant

09 Port Washington
Grafton
Muskegon

MICHIGAN

Madison
Milwaukee
Grand Rapids
Lansing

WISCONSIN
10
02
06
Holland

12
05 Racine
Kalamazoo

Monroe
Delavan
04
Kenosha
Jackson

Dubuque
07
Coldwater

Freeport
Chicago
Sturgis

IOWA
08
01
Bryan

Clinton
Dixon
Joliet

Davenport
55 miles (90km)
Fort Wayne

Kewanee
La Salle
03
Kankakee
OHIO

Galesburg
ILLINOIS
110 miles (175km)
Kokomo
Portland

Peoria
165 miles (265km)

Beardstown
Bloomington
Lafayette
INDIANA
Muncie
Lebanon
Richmond

Springfield
Champaign
Decatur
11 Indianapolis

*Smack dab between Wisconsin and Illinois on Lake Michigan, Chicago has plenty of
Great Lakes fun within reach. Sample the region's cheeses, hike massive dunes, learn
the history of Harley Davidson and see some of Frank Lloyd Wright's best works.*

CHICAGO

ARTS & CULTURE   HISTORY   OUTDOORS   FOOD & DRINK   FESTIVALS & EVENTS   MUSIC & FILM

——— ONE HOUR FROM ———

01 **Indiana Dunes National Lakeshore**
At the southern tip of Lake Michigan, 15 miles of white-sand dunes and more than 50 miles of trails await outdoor adventurers. Swim, bike, fish, ski or hike, depending on the season – or camp overnight from April to October. Beaches fill up fast in the summer, so arrive early to pick the best spot. Less confident swimmers can head to West Beach, the only one with a lifeguard station. *www.nps.gov/indu; 1hr 10min by car, 1hr 30min by South Shore Line from Millennium Station.*

——— TWO HOURS FROM ———

02 **Harley Davidson, Milwaukee, Wisconsin**
No matter how you feel about motorcycles, Harley Davidson ranks among the most iconic American companies. Delve into its storied past in Milwaukee, where it was founded by William S Harley and Arthur Davidson in 1901, by visiting the plant and museum, where you can hop on a Harley yourself. *www.harley-davidson.com/us/en/museum. html; 1hr 40min by Hiawatha Service from Union Station, or 2hr 30min by car.*

03 **Big Vermillion River, Illinois**
Yes, you can go white-water rafting in Illinois. Book a trip at Vermillion River Rafting near Oglesby and splash through nine miles of white water on the Big Vermillion River. Suitable for beginners, the river contains rapids up to Class III (at least, when the water is high in the summer). *www.vermillionriverrafting.com; 2hr by car.*

04 **Lake Geneva Shore Path, Wisconsin**
You'll find something fascinating along just about any stretch of this nearly 26-mile path, which was originally forged to link Native American villages. It winds past excellent lake views, wooded stretches and beautiful estates. Strike out in either direction from Lake Geneva Library for the easiest route. *www.visitlakegeneva.com/ lake-geneva-shore-path; 2hr by car.*

05 **Racine, Wisconsin**
Architecture aficionados shouldn't miss Racine, which is home to several notable Frank Lloyd Wright buildings. Manufacturer SC Johnson is based here, and its third-generation CEO commissioned Wright to build not only his home, Wingspread, but also the administration building and research tower for SC Johnson itself. Tour both to gain an appreciation of Wright's versatility as a designer. *https://reservations.scjohnson.com; 2hr 20min by car.*

06 **Windmill Island Gardens, Holland, Michigan**
Get a taste of Dutch life in delightfully kitschy Windmill Island Gardens. The top attraction in the aptly named town of Holland, this 36-acre park contains gardens, dykes, canals, picnic areas and, of course, a giant windmill imported from the Netherlands in 1964. In the spring, more than 100,000 blooming tulips draw admiring crowds to town. *www.cityofholland.com/ windmillislandgardens; 2hr 20min by car.*

FRANK LLOYD WRIGHT TRAILS

A native of Wisconsin, Frank Lloyd Wright designed more than 20 buildings in the region that are open to the public, and state-organised trails make it easier to find them. Wisconsin's trail (*www.wrightin wisconsin.org*) goes from Racine to Richland Center, crossing the prairies that inspired his famed Prairie School, and showcasing nine designs, including Wright's own home of Taliesin. In Illinois (*www.enjoyillinois. com/history/ frank-lloyd-wright-trail*), 13 buildings are linked over more than 300 miles, beginning with Wright's home and studio in Oak Park (*www. flwright.org*).

© Erik Tanghe / Shutterstock. © PVstock.com / Alamy Stock Photo; © Mint Images / Getty Images

07 **Apple Barn Orchard & Winery, Elkhorn, Wisconsin**

Apples, strawberries and pumpkins are on the pick-your-own schedule here – perfect for keeping kids busy while adults sample well-crafted fruit wines. Pies and cider donuts are on offer, as well as a maze and tractor rides in the fall.
www.applebarnorchardandwinery.com; 2hr 30min by car.

——— THREE HOURS FROM ———

08 **Raven's Grin Inn, Mt Carroll, Illinois**
Chicago's quirkiest day trip just might be this family-run haunted house in small-town Illinois. The 1870 mansion has been both a brothel and a car dealership, and local lore says it's actually haunted. Daily tours, which last nearly two hours and include going down two slides, are interactive, and humorous stories are served up alongside scares. It's hard to say more without spoiling the fun – this is an attraction that has to be experienced.
www.hauntedravensgrin.com; 2hr 40min by car.

09 **Paramount Plaza, Grafton, Wisconsin**

See the spot where a small record label – started by a Wisconsin chair company in 1917 –pulled together some of the most legendary blues and jazz artists of all time. Jelly Roll Morton, Charley Patton, Son House and Louis Armstrong all recorded tracks at Paramount Records in Grafton. A self-guided walking tour starts from this commemorative plaza, which features statues of some of the greats and a

pavement painted like a piano keyboard.
www.travelwisconsin.com/history-heritage/paramount-plaza-walking-tour-290815; 2hr 50min by car.

10 **Dane County Farmers Market, Madison, Wisconsin**

There's much to love about Madison, but a trip to its farmers market – one of the largest in the nation – is a must-do. Every Saturday morning, food vendors of every stripe line up around the elegant white marble state capitol to sell their wares. This being Wisconsin, you should seek out some squeaky, fresh cheese curds.
www.dcfm.org; 3hr by car.

© youngryand / Shutterstock

© Shutterstock / HodagMedia

11 Bottleworks District, Indianapolis, Indiana

One of the most visible signs of Indianapolis' revitalised core is this massive new urban development, located on the 12-acre campus of a former Coca-Cola bottling plant. Cruise the 30,000-sq-ft food hall, filled with local vendors and framed by art deco terracotta facades, see a movie at the Living Room Theater and cap it all off with a cocktail at the rooftop bar.
www.bottleworksdistrict.com; 3hr by car.

12 New Glarus Brewing, Wisconsin

In a state filled with breweries, New Glarus Brewing manages to stand out. Its beer is only sold in Wisconsin, and it's not uncommon for out-of-state visitors to load up on cases of Spotted Cow, Uff-da or Bubbler (Wisconsin's word for a water fountain). Stop in for a tour, then relax on the outdoor patio. Bonus: the town of New Glarus was founded by Swiss immigrants and is full of alpine charm; sleep in a chalet or sample fondue.
www.newglarusbrewing.com; 3hr by car.

INDY 500

More than 250,000 people flock to Indianapolis (3hr by car) on Memorial Day weekend (in late May) for the Indianapolis 500 (pictured opposite), an open-wheel race that is the largest spectator sporting event in the world. But the fun isn't limited to one weekend – or to the famous Brickyard (*www. indianapolis motorspeedway. com*), as the speedway is nicknamed. 'The Month of May', as it's known to fans and Indy residents, is a four-week, citywide party that includes festivals, a 5K, a Grand Prix and a massive parade through the heart of downtown.

05

US COLLEGE TOWN GETAWAYS

Let's put the sleepy college town stereotype to rest. These days, even the smallest of schools boasts a standout museum or two and a lively food and drink scene and, oh yeah, sports.

Ann Arbor, Michigan

Catch culture in the home of the University of Michigan. Hear live folk music at The Ark (www.theark. org), dig into the past at Kelsey Museum of Archaeology (www. lsa.umich.edu/kelsey) and see the ever-changing art at Graffiti Alley – then enjoy a corned beef sandwich at world-famous Zingerman's Deli (www. zingermansdeli.com). *1hr from Detroit by car.*

Hanover, New Hampshire

Once you've been to Hanover you might not want to leave – Dartmouth's home base is frequently ranked among the best places to live in the US. Museums are the star attractions here: the Hood Museum of Art (https://hoodmuseum. dartmouth.edu) has been acquiring arte-facts since 1772. *2hr from Boston by car.*

Boulder, Colorado

Boulder prides itself on near-perfect weather and a manicured campus. But it's the city's dedication to pairing plenty of green space with cultural outlets, like gourmet restaurants and energetic bars, that makes it special. *30min from Denver by car.*

Flagstaff, Arizona

The laid-back charms of the home of Northern Arizona Uni-versity are many, from a pedestrian-friendly historic downtown, bedecked with vintage neon, to hiking and skiing in the country's largest ponderosa pine forest. *2hr 30min from Phoe-nix by car.*

Oxford, Mississippi

Alongside tailgating, eating and drinking are the pillars of the Oxford experience. Try 'haute Southern' treats like Royal Red Shrimp Mac & Cheese from one of the excellent restaurants helmed by award-winning chef John Currence – you'll love (www.citygrocery-online.com/snackbar). *1hr 30min from Memphis by car.*

Chapel Hill, North Carolina

Bursting with energy, Chapel Hill has a standout food and drink scene surround-ing the UNC campus. Lantern (www.lantern-restaurant.com), an award-winning Asian fusion restaurant, is one of the best restaurants in the state, and the cosy, wood-panelled Crunkleton (www. thecrunkleton.com) is a cocktail-lover's delight. *2hr from Charlotte by car.*

Waco, Texas

Waco's Magnolia Market at the Silos (www.magnolia.com/silos) draws more visitors than the Alamo. Once you've shopped, played and eaten at 'Fixer-Upper' duo Chip and Joanna Gaines' biggest renovation project, stroll Baylor's 1000-acre campus or stand-up paddle straight through town on the Brazos River. *1hr 30min from Dallas by car.*

State College, Pennsylvania

There's a reason this region is nicknamed 'Happy Valley' – State College has a cheerful collection of bars and plenty of hikes and mountain-biking trails in the surrounding Allegheny Mountains. In summer, catch the sophisticated Central Pennsylvania Festival of the Arts (https://arts-festival.com). *3hr from Philadelphia by car.*

New Haven, Connecticut

Yale's campus offers a wealth of attractions, but there's another side to New Haven. Far from elite Ivy League, the city's well-aged dive bars, ethnic restau-rants, barbecue shacks and cocktail lounges make for a lively but ego-free atmosphere. *1hr 30min from NYC by car.*

Ames, Iowa

For a small city, Ames has a huge green foot-print. The city is home to 35 leafy parks, more than 30 miles of bike and pedestrian trails, a water park and five golf courses. Throw in a couple of breweries, a farmers market and a winery and you've got yourself a weekend. *30min from Des Moines by car.*

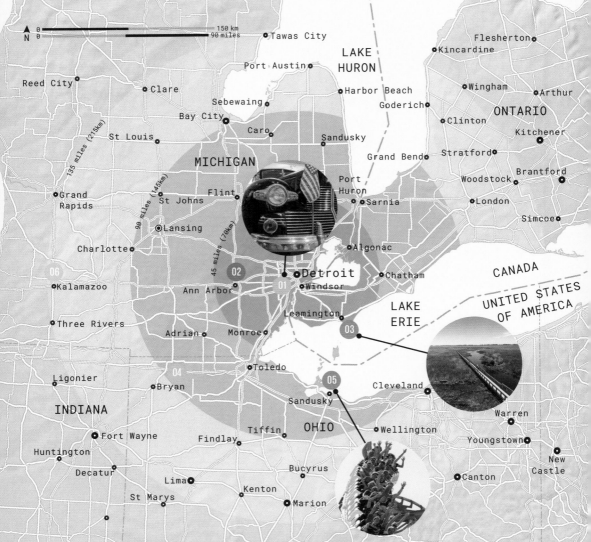

MICHIGAN

N

0 ——— 150 km
0 ——— 90 miles

Tawas City

LAKE HURON

Port Austin

Flesherton

Kincardine

Reed City

Clare

Harbor Beach

Goderich

Wingham

Arthur

ONTARIO

Sebewaing

Bay City

Caro

Sandusky

Clinton

Kitchener

St Louis

Grand Bend

Stratford

Simcoe

Charlotte

Grand Rapids

St Johns

Flint

Port Huron

Sarnia

Woodstock

London

Kalamazoo

Lansing

02

01

Detroit

Windsor

Algonac

Chatham

Brantford

135 miles (215km)

90 miles (145km)

45 miles (70km)

Ann Arbor

Three Rivers

Adrian

Monroe

Leamington

03

LAKE ERIE

CANADA

UNITED STATES OF AMERICA

Ligonier

04

Bryan

Toledo

05

Sandusky

Cleveland

Warren

INDIANA

Fort Wayne

Tiffin

OHIO

Wellington

Youngstown

New Castle

Huntington

Findlay

Bucyrus

Canton

Decatur

Lima

Kenton

Marion

St Marys

06

Motor out of Motor City to take in the state's many charms or plunge into fun at an amusement park or a national park. There's even a train line for the car-free.

DETROIT

● ARTS & CULTURE   ○ HISTORY   ● OUTDOORS   ○ FOOD & DRINK   ○ FESTIVALS & EVENTS   ○ MUSIC & FILM

──── ONE HOUR FROM ────

01 **Henry Ford Museum & Greenfield Village, Dearborn, Michigan**

Plunge into American history, inside and out. The indoor Henry Ford Museum contains a wealth of American culture, such as the chair Lincoln was sitting in when he was assassinated, the limo in which Kennedy was killed and the bus on which Rosa Parks refused to give up her seat. The adjacent outdoor Greenfield Village features Thomas Edison's laboratory and the Wright Brothers' aeroplane workshop. *www.thehenryford.org; 30min by car.*

02 **Ann Arbor Art Fair, Ann Arbor, Michigan**

While most associate the walkable and bookish Ann Arbor with the University of Michigan and its football team, there are many who think 'art fair' when they consider this city. For four days each, over 1000 artists display their works outside, along 30 city blocks, in America's largest juried art fair. *www. visitannarbor.org; mid-July; 1hr by train.*

03 **Point Pelee National Park, Ontario, Canada**

On a peninsula jutting into Lake Erie, at Canada's southernmost mainland point, this national park features nature trails, a marsh boardwalk, forests, lovely sandy beaches and seasonal treats. The autumn migration of monarch butterflies is a spectacle of swirling black and orange. Myriad transiting bird species lure spotters in spring and autumn. And in winter, the silence amidst the trees is awesome. *www.pc.gc.ca; 1hr20min by car.*

──── TWO HOURS FROM ────

04 **Historic Sauder Village, Archbold, Ohio**

See how hardy settlers turned Ohio's Great Black Swamp from an ice-age-era bog into some of the Midwest's most productive farmland. Preserved buildings spanning the years 1803 to 1928 show what life was like before the modern age. Costumed guides demonstrate how everyday items were fashioned from glass, metal, wood and clay. You can tour an old schoolhouse, farmhouse, general store and meet the farm animals, before tucking into hearty period dishes in the restaurant. *www.saudervillage.org; May-Oct; 2hr by car.*

05 **Cedar Point, Sandusky, Ohio**

Cedar Point on Lake Erie is one of the world's top amusement parks, known for its 17 adrenaline-pumping rollercoasters. Stomach-droppers include the Top Thrill Dragster, among the globe's tallest and fastest rides. It climbs 420ft into the air before plunging and whipping around at 120mph. The Valravn is the world's longest 'dive' coaster, dropping riders at a 90-degree angle for 214ft. *www. cedarpoint.com; 2hr 30min by car.*

──── THREE HOURS FROM ────

06 **Breweries, Kalamazoo, Michigan**

Kalamazoo has an offbeat charm that will surprise first-time visitors. But it's the local beer that has got people talking; over a dozen breweries produce a huge range of them. The leader is Bell's, one of the top microbreweries in the country. *www.discoverkalamazoo.com; 2hr 40min by train.*

THE LOCAL'S VIEW

'I love the grit and never-say-die spirit of Detroit, but every so often I get wanderlust for quaint little towns. Luckily, there are a bevy of small towns to escape to across Michigan. In one hour's drive, I can be in downtown Chelsea, a cute little town with a surprising bit of 1890s culture, or lakeside Holly, an old-timey burg that really gets into the holiday spirit with all sorts of Christmas displays. Or sometimes, I just pick an inviting two-laner and see where it takes me.'

Sarah Ulicny, editor

© Danita Delimont / Alamy Stock Photo; © Chiyacat / Shutterstock; © Denise Panyik-Dale / Getty Images

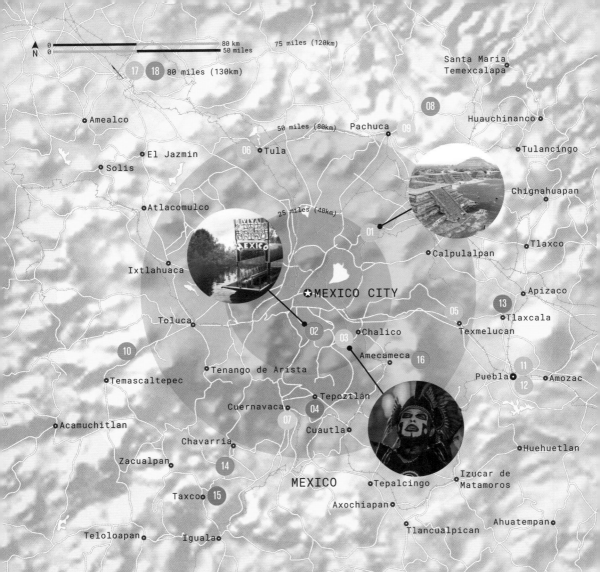

N

0   80 km
0   50 miles

75 miles (120km)

17 18 80 miles (130km)

Santa Maria
Temexcalapa

• Amealco

08

Huauchinanco

• El Jazmin

• Solis

06 Tula

• Atlacomulco

50 miles (80km) Pachuca
09

• Tulancingo

Chignahuapan

25 miles (40km)

01

Calpulalpan

• Tlaxco

• Ixtlahuaca

MEXICO CITY

• Apizaco

• Toluca

02

Chalico

03 Amecameca
16

05 13 Tlaxcala
Texmelucan

10

• Tenango de Arista

Puebla 11
12 • Amozac

• Temascaltepec

Cuernavaca

• Tepoztlán
07 04
Cuautla •

• Acamuchitlan

• Chavarria

Zacualpan 14

MEXICO

• Huehuetlan

Taxco • 15

Tepalcingo •

Izucar de
Matamoros

• Teloloapan

Iguala •

Axochiapan •

Ahuatempan •

Tlancualpican •

Climb to the top of an ancient pyramid, sip mezcal under the stars, float down a canal on a wooden boat while a mariachi band plays: there's no shortage of things to see, do, and taste around Mexico City's sprawling metropolis.

MEXICO CITY

● ARTS & CULTURE ● HISTORY ● OUTDOORS ● FOOD & DRINK ● FESTIVALS & EVENTS ● MUSIC & FILM

——— ONE HOUR FROM ———

01 Pyramids of Teotihuacán
The ruins of the ancient pre-Columbian city of Teotihuacán, complete with a pair of massive pyramids and what's left of the once-opulent Temple of Quetzalcoatl, are a wonder to behold. Walk down the Avenue of the Dead and climb to the heights of the Pyramid of the Moon and the Pyramid of the Sun for sweeping views over the site. *1hr by car or 1hr 20min by bus.*

02 Boat ride, Xochimilco
Technically, you're still in Mexico City (in the deep south of it). But it won't feel like it as you float down a forested waterway on a colourfully painted boat, with a live mariachi band drifting past and locals sharing tacos and tequila while enjoying the ride. This vast canal system was built by the Aztecs; on weekends especially, it's still a popular escape from the city. *1hr by car or 1hr 20min by bus.*

03 Day of the Dead, Mixquic
Now's your chance to experience the unique el Dia de los Muertos (Day of the Dead) festival in person. Mixquic, in Tláhuac, to the southeast of Mexico City, is widely considered one of the best places to celebrate on 2 November. Expect carnival-style street-food vendors, traditional dancing, special exhibits, a cemetery brimming with fresh flowers, and friendly locals in costume. *2 Nov; 1hr 20min by car or 2hr 40min by bus.*

04 Tepoztlán
Sedona is to the USA what Tepoztlán is to Mexico: a place known for its creative and spiritual energy (and a magnet for New Age travellers). According to legend, it's the birthplace of Quetzalcóatl, serpent god of the Aztecs. Today, Tepoztlán remains a centre for Náhuatl culture, and its historic heart, surrounded by tall cliffs and featuring charming restaurants and an indigenous crafts market, is a delight. *1hr 30min by car or 2hr by bus.*

05 Ex-Hacienda de Chautla
On the road to Puebla from Mexico City, the Ex-Hacienda de Chautla makes for a fascinating pit stop. The 18th-century ranch features an English-style villa called El Castillo ('the Castle') and the first hydroelectric dam in Latin America. Once privately owned, it's now open to the public – you're free to wander the grounds, which include a small lake and a courtyard with a historic fountain. *1hr 30min by car.*

06 Tula
Located in the Tula Valley in the state of Hidalgo, Tula is the ancient capital of the Toltec of central Mexico. The archeological site offers clues to the culture that thrived here between AD 850 and 1150. The major highlight is the Pyramid of Quetzalcoatl, topped by a quartet of massive warrior figures carved in stone. *www.inah.gob.mx/zonas/80-zona-arqueologica-y-museo-de-sitio-de-tula; 1hr 30min by car.*

——— TWO HOURS FROM ———

07 Anticavilla Hotel & Spa, Cuernavaca
This dreamy hotel and spa is the perfect

MEGACITY MOVES

Both in terms of size and population, Mexico City is gigantic (bigger than New York City, for comparison). Several of the destinations listed as 'getaways' in this book are technically within the city limits – notably, Xochimilco, known for its system of canals plied by traditional wooden boats – they're just so far out and they feel like another place. Take advantage of the STC (Mexico City Metro) and the modern light rail to get where you need to go.

weekend retreat. Established in a former convent in Cuernavaca ('the city of eternal spring'), Anticavilla has the best of the old world – romantic stone architecture, gardens blooming with vibrant flowers and foliage – and the new, with twelve luxurious suites, a gourmet open-air restaurant, a mezcal bar, and a swimming pool fringed with palm trees. *www.anticavillahotel.com; 1hr 40min by car.*

08 **Basaltic Prisms of Santa María Regla**
One of Mexico's natural wonders, you'll have to see this sight to belive it – a striking ravine and waterfalls surrounded by towering rock columns that were formed by the cooling of volcanic lava. Luckily, there's a wonderful hotel, the historic Hacienda Santa María Regla, right next to the ravine. If you're only day-tripping, take advantage of the hacienda's good restaurant. *www.haciendaderegla. com.mx; 2hr by car.*

09 **Real del Monte**
In the mountain town of Real del Monte (also known as Mineral del Monte), you'll find museums dedicated to the region's mining tradition. As it's in a different state – Hidalgo – there's also a different regional cuisine to sample. The town's kitchens specialise in tamales and, thanks to the English miners who came to work here in the 19th century, Cornish-style pasties (similar to empanadas) stuffed with meat and potatoes. *2hr by car.*

10 **Piedra Herrada Sanctuary**
The monarch butterfly reserve at Piedra Herrada – an easy two-hour drive

© Noradoa / Shutterstock

10

from Mexico City along the state highway – is a Unesco World Heritage Site. The sanctuary itself is a short but relatively challenging hike through the woods from the site's entrance. Aim to visit between November and March, when masses of butterflies migrate from Canada to Mexico, to witness them fluttering between flowers and tree branches all around you. *www.rutamonarca.org; 2hr by car.*

11 **Puebla**
Low-key Puebla is known to Mexicans for its beautifully preserved historic centre, where hand-painted Talavera tiles add a distinctive flair to many buildings. Also be sure to check out the central square (the zócalo) with its stately cathedral, first-class museums displaying pre-Hispanic and colonial art and artefacts, and the city's lively dining and drinking scene. *2hr by car or 2hr 20min by bus from TAPO station.*

12 Biblioteca Palafoxiana, Puebla

Founded in 1646, Biblioteca Palafoxiana in Puebla's historic centre is widely considered the first library in the Americas. It was certainly the first in colonial Mexico, and the library still has an extensive collection of more than 45,000 books that date as far back as the 15th century. *www.puebla.travel/es/museos/item/ biblioteca-palafoxiana; 2hr by car or 2hr 20min by bus from TAPO station.*

13 Tlaxcala museums

The small city of Tlaxcala excels at cultural preservation: many of the area's most historic buildings have been transformed into public museums. Learn about regional artisan traditions at the Museo Vivo de Artes y Tradiciones Populares, see bullfighting accoutrements at the Museo Taurino, trace central Mexican history at the Tlaxcala Regional Museum, and enjoy little-known Frida Kahlo paintings at the Art Museum of Tlaxcala. *2hr 15min by car.*

14 Cacahuamilpa Caverns

One of Mexico's most breathtaking natural sights, this cave complex provides an unmissable underground adventure. The Grutas de Cacahuamilpa features easily navigable pathways, cathedral-like chambers, and ethereal stalactites and stalagmites. You'll need a guide for the cave tour (provided with admission) but not for the short hike to the natural pools at nearby Río Dos Bocas. Try to avoid weekends, when the caves get extra busy. *www.cacahuamilpa.conanp.gob.mx; 2hr 30min by car.*

TRAVELLING BY TAXI

It's important to note that in Mexico City you can't just flag a taxi as you would in many other cities. For safety reasons, you should always find a taxi at an official taxi stand, or call for one. This detail makes getting out of the city a little trickier, too: in order to get to one of the four bus stations that serve destinations outside of CDMX, take the metro, or ask the front desk at your hotel or hostel to call a taxi for you.

14

© Roberto Michel / Shutterstock

16

THREE HOURS FROM

15 **Taxco**

A hotspot for silver mining since the 16th century, the picturesque city of Taxco, in Mexico's Central Highlands, doesn't have much silver left these days. It still draws in visitors thanks to its stunning mountain location and attractive historic centre, with the twin towers of Templo de Santa Prisca rising up from a cluster of white buildings. Preservation laws insist new construction projects match the colonial cityscape in scale, style, and materials. *2hr 50min by car.*

16 **Izta-Popo Zoquiapan National Park**

A high-altitude national park known for its craggy mountain summits, active snow-capped volcanoes, and tropical birdlife, Izta-Popo is a breath of fresh air for city-weary travellers. Go for a hike along Paso de Cortés pass and keep your eyes peeled for monkeys and the endangered volcano rabbit. The nearby village of El Silencio is a great spot for lunch, or you can pack a picnic. *3hr by car.*

15

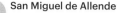

17 San Miguel de Allende

Unesco named the city of San Miguel de Allende a World Heritage Site in 2008. It's a colonial gem with a cosmopolitan spirit: historic landmarks like the 17th-century neo-Gothic cathedral La Parroquia sit alongside stylish cafes and bars. Get the best of both worlds by checking in to the Rosewood San Miguel de Allende, a luxurious hotel designed and crafted by local artisans. *www.rosewoodhotels.com/en/san-miguel-de-allende; 3hr 20min by car.*

18 El Charco del Ingenio

Just outside San Miguel de Allende, Jardín Botánico El Charco del Ingenio is a natural reserve, bird sanctuary and botanical garden with a large collection of cacti and endangered desert plants. Follow paths leading through the gardens, observing exotic birds along the way, heading towards the bottom of the canyon and the freshwater spring that gives the gardens their name. *www.elcharco.org.mx; 3hr 20min by car.*

**NORTH,
SOUTH,
EAST, WEST**

You can catch a bus to almost anywhere in Mexico from the capital city. The trick is to know which of the four bus terminals you need to start from. Generally speaking, if you're headed somewhere north of Mexico City, go to Terminal Central del Norte (accessible on the yellow and green metro lines), and if you're headed south, go to Terminal Central del Sur (pink and blue metro lines). Heading east, you need Terminal de Autobuses de Pasejeros del Oriente (pink and green metro), and travelling somewhere west of CDMX, use Terminal Centro del Poniente (pink metro).

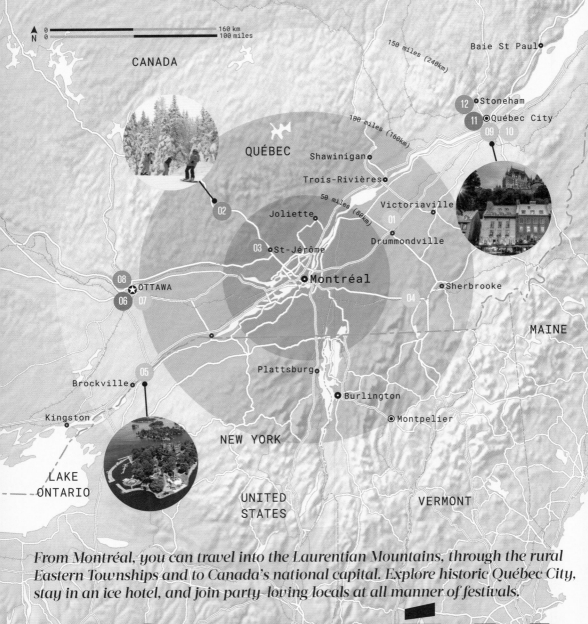

N

0 ————— 160 km
0 ————— 100 miles

CANADA

QUÉBEC

150 miles (240km)

Baie St Paul

12 • Stoneham

11 ◉ Québec City

09 10

100 miles (160km)

Shawinigan

Trois-Rivières

50 miles (80km)

Victoriaville

Joliette

01

Drummondville

03 • St-Jérôme

Montréal

Sherbrooke

08 • OTTAWA

06 07

04

MAINE

05

Brockville

Plattsburg

Kingston

Burlington

◉ Montpelier

NEW YORK

LAKE
ONTARIO

UNITED
STATES

VERMONT

From Montréal, you can travel into the Laurentian Mountains, through the rural
Eastern Townships and to Canada's national capital. Explore historic Québec City,
stay in an ice hotel, and join party-loving locals at all manner of festivals.

MONTRÉAL

● ARTS & CULTURE   ● HISTORY   ● OUTDOORS   ● FOOD & DRINK   ● FESTIVALS & EVENTS   ● MUSIC & FILM

——— ONE HOUR FROM ———

01 Drummondville

While many competing legends surround the Québécois concoction of French fries, brown gravy and fresh cheese curds, Drummondville claims poutine as its own; a local restaurateur contends that he invented the dish in the 1960s. The town celebrates this 'mess' of a meal at the annual Festival de la Poutine, an August weekend of concerts and food events. *www.festivaldelapoutine.com; Aug; 1hr 30min by car or train.*

02 Mont-Tremblant

Skiers, snowboarders and other winter-sports enthusiasts head for Québec's largest snow resort in the Laurentian Mountains northwest of Montréal. Mont-Tremblant's 14 lifts and 100+ trails let skiers and boarders zoom over more than 300 hectares of terrain. Off the slopes, try dog sledding, ice climbing, ice fishing or tubing, or relax at the spa. *www.tremblant.ca; 1hr 30min by bus from Montréal-Trudeau International Airport or 1hr 50min by car.*

03 La Route Verte

La Route Verte is a network of more than 5000km of cycling paths throughout Québec. One popular pathway, Le P'tit Train du Nord, takes you between Mont-Tremblant and the Montréal region, with auberges to stay in along the way. A bus service for cyclists can transport your luggage or shuttle you to your starting point; it rents bikes, too. *www.routeverte.com, www.laurentides .com/en/linearpark, www.autobuslepetit traindunord.com; 1hr 30min by car or bus.*

04 Eastern Townships

In the agricultural communities from Lac-Brome to Sutton to Magog in Québec's Cantons-des-L'Est (Eastern Townships), you'll find cheesemakers, berry farms, chocolate shops, cideries, microbreweries, wineries and farm-to-table dining rooms, plus country inns where you can spend the night. Spring is maple syrup-making season, when you can visit local 'sugar shacks'. In autumn, you'll add brilliant fall colours to your eating and drinking itinerary. *1hr 30min by car.*

05 St Lawrence Region

Several sights along the St Lawrence River delve into eastern Canada's 19th-century roots. Visit Upper Canada Village, which recreates life in the 1860s. Tour Fulford Place, the grand manor that George Fulford constructed with the fortune he made hawking 'health' drugs. Catch a ferry to 120-room Boldt Castle, on one of the Thousand Islands, built by the owner of New York's Waldorf Astoria Hotel. *1hr 30min by car.*

——— TWO HOURS FROM ———

06 Ottawa

You could spend days exploring the grand museums in Canada's national capital. Top spots include the National Gallery of Canada, home to the country's largest visual arts collection, and the Canadian Museum of History, where you can walk through centuries of social and cultural change. The sobering Canadian War Museum and kid-friendly Canadian Museum of Nature are worth seeing, too. *2hr by car or by train.*

QUÉBEC'S CHEESE TRAIL

Do you love fromage? Québec produces more than 500 varieties of cheese, making it an excellent place for a tasting tour. Several regions near Montréal, notably the Eastern Townships, the Laurentians, Lanaudière and Outaouais, are hubs of cheese production. To organise your own cheese getaway, refer to Fromages d'Ici (www. fromagesdici. com), which offers detailed descriptions of the cheeses made in Québec and where to find them. Terroir et Saveurs (www. terroiretsaveurs. com), a Québec agritourism organisation, provides a map of Québec cheesemakers, Route des Fromages Fins du Québec.

© Cybernesco / Getty Images; © grandriver / Getty Images; © Matt Champlin / Getty Images

07 RBC Ottawa Bluesfest

One of the many summer festivals in Canada's capital city, BluesFest brings ten days of concerts to Ottawa's LeBreton Flats Park every July. From local musicians to big names (past performers have included Bryan Adams, Lady Gaga and Dave Matthews Band), the musical styles start with the blues, but encompass jazz, rap, world music and other genres, too. *www.ottawabluesfest.ca; Jul; 2hr by car or train.*

08 Gatineau Park

Outside of Ottawa on the Québec side of the Ottawa River, massive Gatineau Park covers 361 sq km in the Gatineau Hills. Go hiking, cycling, canoeing, and picnicking in the warmer months; in winter, follow the trails on snowshoes or cross-country skis. Nearby, you can soak and steam in the pools and saunas at Nordik Spa-Nature, North America's largest Nordic-style spa. *www.ncc-ccn.gc.ca/places-to-visit/ gatineau-park; 2hr by car.*

—— THREE HOURS FROM ——

09 Québec City

Wander the narrow streets of Québec's old city, then visit Musée de la Civilisation to trace Canada's early history. Explore the park-like Plains of Abraham, where a 1759 battle between British and French troops helped shape modern North America. Take time to tour Musée National des Beaux-arts du Québec, too, where the artworks date from ancient times to the present day. *3hr by car or by train.*

10 Winter Carnival, Québec City

Yes, it's cold in February, yet Québec City braces itself for loads of snowy-weather fun at the annual Winter Carnival. Watch the ice canoe race on the frozen St Lawrence River, check out the ice palace and snow sculptures, and toast the season with a drink of caribou, a spiced punch made from red wine and hard liquor. *www.carnaval.qc.ca; Feb; 3hr by car or by train.*

© Vlad G / Shutterstock

© William P. McElligott

Wendake

Learn about the culture of Québec's Wendake First Nation at the Huron-Wendat Museum, part of the modern Hôtel-Musée Premières Nations complex. Time your visit to take in a fireside storytelling session in a traditional longhouse. Nearby, tour the reserve's historic Notre-Dame-de-Lorette Church. From late May through early October, Tourism Wendake transports visitors by shuttle from central Québec City. *3hr by car or by train.*

 Hôtel de Glace

Every winter, a new Ice Hotel – the only one in North America – is constructed near Québec City, welcoming guests for chilled cocktails and a night spent in a frosty palace, built entirely of snow and ice. If sleeping on a fur-covered ice block doesn't appeal, join a 45-minute tour of the property by day. The Ice Hotel normally opens from late December to March.
www.hoteldeglace-canada.com; 3hr by car or by train.

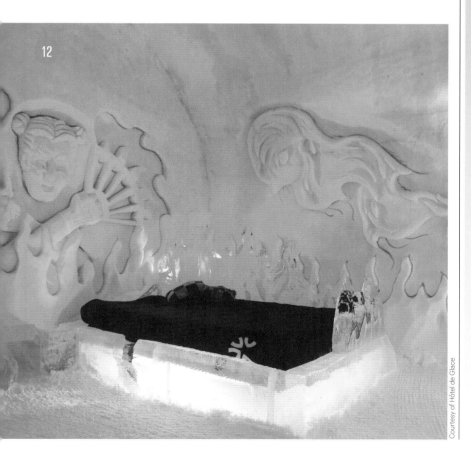

Courtesy of Hôtel de Glace

TOP 10 NORTH AMERICAN SKI ESCAPES

Aspen, Colorado

Iconic, outrageously posh and overwhelmingly beautiful, Aspen has it all. America's best powder run can be found at 12,000ft in Highlands Bowl. Buttermilk has perfect groomers for families, while Aspen Mountain (Ajax) goes right into town. *Nov-Apr (plus summer lift service); lift tickets from $169; 4hr from Denver by car, bus or shuttle.*

Copper, Colorado

Copper isn't as glitzy as some other Colorado resorts, but offers up amazing terrain, a few super steep bowls and fun hike-to runs. Located in Summit County, the resort and little village gives easy access to Vail, Keystone, Breckenridge and Arapahoe Basin. *Nov-Apr (plus summer lift service); lift tickets from $150; 2hr from Denver by car, bus or shuttle.*

Crested Butte, Colorado

This is the stuff of legend: steep terrain, historic town centre, down-to-earth locals and some of the best extreme skiing on the planet. While best suited to a long weekend trip, the awesome terrain and laid-back air make this one of Colorado's top ski destinations. *Nov-Apr (plus summer lift service); lift tickets from $111; 4hr from Denver by car, bus or shuttle.*

Taos, New Mexico

With over 300in of the fresh stuff, limited crowds and easy access to sophisticated Santa Fe, Taos is a small-but-mighty 1200-acre resort, cut from the independent spirit of the American Southwest. *Nov-Apr (plus summer lift service); lift tickets from $105; 2hr from Santa Fe by car, bus or shuttle.*

Snowbird, Utah

As you ride the Snowbird Aerial Tram to the 11,000ft summit of Hidden Peak, the grandeur and immensity of this powder-keg resort reveals itself. Spend a week finding new lines in the 2500 acres of world-class terrain, heading out to other nearby resorts, and just enjoying the village's fun restaurants and activities. *Nov-May (plus summer lift service); lift tickets from $119; 40min from Salt Lake City by car, bus or shuttle.*

Chase champagne powder, ignite the night with high-octane après-ski parties, and live life on the vertical edge with North America's best ski adventures.

Alta, Utah

With transcendent champagne powder, impossibly cool lines and a hard-charging attitude, this is Utah's premier ski resort. From the Wildcat Base, at 8530ft, you get access to over 2200 acres, with a whopping average snowfall of 543in. *Nov-May (plus summer lift service); lift tickets from $139; 40min from Salt Lake City by car, bus or shuttle.*

Kirkwood, California

This mighty resort has the gnarliest lines imaginable, from steep chutes to sweet natural halfpipes and a cornice drop that sends you flying 20ft down. The best part of Kirkwood is its small size – it still feels like a mom-and-pop operation. *Nov-May (plus summer lift service); lift tickets from $86; 1hr 30min from Reno by car, bus or shuttle.*

Whistler Blackcomb, British Columbia

There's nothing small about Canada's twin mountains of Whistler and Blackcomb: this resort has over 8000 acres of skiable terrain (the most in North America) topped with a cool village, plenty of deep snow and beautiful mountain views. *Nov-May (plus summer lift service); lift tickets from $149; 2hr from Vancouver by car, bus or shuttle.*

Crystal, Washington

Crystal Mountain has arguably the best gate-accessed, side-country skiing in North America. With just a 10-minute hike, you'll be skiing fresh tracks all day. While the Washington snow can get a little heavy, there's plenty of it, with around 500in each year. *Nov-May (plus summer lift service); lift tickets from $74; 2hr from Seattle by car, bus or shuttle.*

Stowe, Vermont

Skiing on the US East Coast may pale in comparison to the action out west, but Vermont's Stowe Mountain Resort offers some great riding, with 2360ft of vertical drop, and a luxury New England vibe. *Nov-May (plus summer lift service); lift tickets from $99; 1hr from Burlington by car, bus or shuttle.*

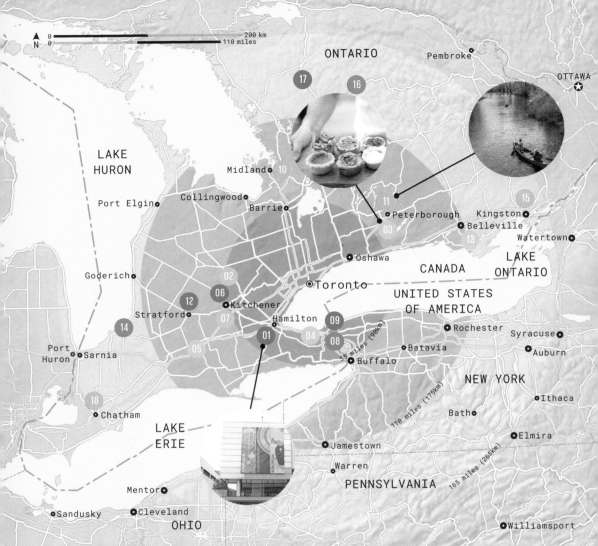

Map labels and markers:

ONTARIO

N

0   200 km
0   110 miles

LAKE HURON

Pembroke

OTTAWA ★

17

16

Midland   10

Collingwood

Barrie

Port Elgin

15

11

Peterborough   Kingston
03   Belleville

13   Watertown

Goderich

Oshawa   CANADA   LAKE ONTARIO

Toronto

02

06   UNITED STATES OF AMERICA

Stratford   12   Kitchener   09   Rochester   Syracuse

14   07   Hamilton   04   08   Auburn

Port Huron   Sarnia   01   05   55 miles (90km)   Batavia

Buffalo   110 miles (175km)   NEW YORK

18   LAKE ERIE   Bath   Ithaca

Chatham   Jamestown   Elmira

165 miles (265km)

Mentor   Warren   PENNSYLVANIA

Sandusky   Cleveland   OHIO   Williamsport

Canada's largest city is the starting point for a diverse range of getaways, from Niagara Falls to the Great Lakes. Whether you're interested in canoeing, beaches, wine tasting, theatre, food trails or local crafts, you'll find it around Toronto.

TORONTO

ARTS & CULTURE ● HISTORY ● OUTDOORS ● FOOD & DRINK ● FESTIVALS & EVENTS ● MUSIC & FILM

——— ONE HOUR FROM ———

01 Hamilton

This industrial city midway between Toronto and Niagara Falls has an emergent visual arts scene, with eclectic contemporary galleries throughout the James Street North district downtown. Time your visit to coincide with the lively Art Crawl, the second Friday each month, or for the September 'SuperCrawl' weekend, when you'll find theatre, music, and dance as well as art events. *www.jamesstreetnorth.ca, www.supercrawl.ca; 1hr by bus, train or car.*

02 Riverfest, Elora

Regularly ranked among Ontario's most scenic small towns for its well-preserved 1800s limestone buildings and pretty setting along the Grand and Irvine Rivers, Elora is also worth visiting in August for its annual music weekend, Riverfest. Musicians from across Canada and beyond, including Blue Rodeo, Serena Rider and Bruce Cockburn, have performed at this outdoor festival held riverside in the town's Bissell Park. *www.riverfestelora.com; Aug; 1hr 30min by car.*

03 Butter Tart Tour

A popular pastry across Ontario (and throughout much of Canada), a butter tart is a single-serving pie oozing with a dark, sugary filling. In the Peterborough region east of Toronto, you can sample these sweets along the Kawarthas Northumberland Butter Tart Tour, a self-guided road trip that entices you to more than 50 bakeries, restaurants and other producers of these gooey confections. *www.buttertarttour.ca; 1hr 30min by car.*

04 Twenty Valley

Known for producing ice wine, Canada's distinctive sweet dessert wine, rural Twenty Valley near Niagara Falls is a growing winemaking district, where riesling, pinot gris, cabernet franc and pinot noir are among the popular varietals. Combine wine touring at the 40+ wineries around the towns of Beamsville, Vineland and Jordan, with a visit to the famous falls or to nearby Niagara-on-the-Lake. *www.20valley.ca; 1hr 30min by car.*

——— TWO HOURS FROM ———

05 Oxford County Cheese Trail

Attention, cheese lovers! This 'cheese trail', west of Toronto, with two dozen stops at cheesemakers, restaurants, and other dairy-related businesses, is for you. There's even a cheese museum, Ingersoll Cheese & Agricultural Museum, which explores the history of Canada's cheese industry, launched in the 1860s; one exhibit recounts an 1866 publicity stunt, when local cheesemakers created a 3300kg 'Mammoth Cheese'. *www.tourismoxford.ca; 1hr 40min by car.*

06 St Jacobs

Old Order Mennonite families still live in the farming communities around St Jacobs, where some sell homemade jams and baked goods at the year-round St Jacobs Farmers Market. Learn more about their culture at The Mennonite Story visitor centre, or take a cycling tour with outfitter Grand Experiences, where you'll lunch with a Mennonite grandmother, visit a harness shop, and wander through a rural general store. *1hr 40min by car.*

THE PARKBUS

To get outdoors from Toronto without a car, check out the Park Bus. This non-profit service provides transportation on select weekends from several Canadian cities to nearby national and provincial parks. In Ontario, the Parkbus runs day trips from downtown Toronto to Elora Gorge, Christie Lake and Rockwood Conservation Area, among other destinations, and weekend excursions to Algonquin, Killarney and Grundy Lake Provincial Parks, the Bruce Peninsula, and Georgian Bay Islands National Park. For details, including schedules and pick-up/drop-off points, see the Park Bus website (*www.parkbus.ca*).

Courtesy of Butter Tart Tour; © Bill Brooks / Alamy Stock Photo; © Alastair Wallace / Getty Images

© Vlad Ghiea / Alamy Stock Photo

 Oktoberfest & Craftoberfest, Kitchener-Waterloo

Many early settlers to Kitchener-Waterloo, now one of Canada's high-tech hotbeds, were of Germanic heritage. K-W's long-running Oktoberfest, one of the largest Bavarian beer festivals outside Munich, celebrates this legacy with music, food and, of course, beer. Launched more recently, the weekend-long Craftoberfest highlights the area's increasing number of craft breweries. Both festivals take place in October. *www.oktoberfest.ca, www.kwcraftoberfest.com; Oct; 1hr 40min by car or by bus.*

 Niagara Falls

One of Canada's most visited attractions, Niagara Falls encompasses three thundering waterfalls: Horseshoe Falls on Ontario's side of the border, and American and Bridal Veil Falls on New York state's shore. Hornblower Niagara Cruise is a highlight; a boat trip beneath the spray. For even more adventure, zipline past the waterfalls as you plunge into Niagara gorge. *1hr 50min by car or 2hr by bus or train.*

09 **Shaw Festival, Niagara-on-the-Lake**

A major North American theatre festival takes place from April to October in the historic town of Niagara-on-the-Lake, north of Niagara Falls. The Shaw Festival produces works by Irish playwright George Bernard Shaw, along with more contemporary plays, on several stages. With two dozen wineries nearby, you can easily combine theatre with a wine-tasting tour. *www.shawfest.com; Apr-Oct; 1hr 50min by car or 2hr 20min by bus or train & bus.*

10 **Sainte-Marie Among the Hurons, Midland**

Beginning in 1639, French Jesuits lived and worked among the indigenous Wendat people at Sainte-Marie Among the Hurons, Ontario's earliest European settlement. Today, you can go back to the 17th century at this historical village, outside the present-day town of Midland. Costumed interpreters help visitors imagine life during this era from the European and indigenous perspectives, as you wander through the two dozen reconstructed buildings. *www.saintemarie amongthehurons.on.ca; 2hr by car.*

11 **Peterborough**

Learn the importance of the canoe to Canada's indigenous people and early settlers at the well-designed Canadian Canoe Museum in Peterborough, at the centre of the lake-filled Kawarthas region. Then try a unique adventure at Lock 21 on the Trent-Severn Waterway, where you can paddle a voyageur canoe into the world's highest hydraulic lift lock and ascend nearly 20m as the lock chamber fills. *www.canoemuseum.ca, www.thekawarthas.ca; 2hr by car or 2hr 30min by train + bus.*

12 Stratford Festival
Started as a Shakespeare fest in 1953, the Stratford Festival is now North America's largest classical repertory theatre. Staged from April to October, productions range from plays by the Bard to works by contemporary playwrights. Set like its British namesake on the Avon River, the town is home to the professional Stratford Chef School, which means plenty of top-notch restaurants too. *www.stratfordfestival.ca; Apr-Oct; 2hr 20min by train, bus or car.*

13 Prince Edward County
Make a weekend of good drink and local food in Prince Edward County, an agricultural district on Lake Ontario, where the appetising attractions include three dozen wineries, a number of craft-beer makers and excellent restaurants. Between dining and drinking stops, follow the County's self-guided 'Arts Trail' to studios and galleries, or chill in the lakeside sand dunes at Sandbanks Provincial Park. *2hr 30min by car.*

——— THREE HOURS FROM ———

14 Pinery Provincial Park
Along the shore of Lake Huron, one of North America's five Great Lakes, this popular provincial park has a 10km-long, dune-backed beach. You can camp near the sand (the park has 1000 campsites), kayak a nearby river channel, or take short hikes through the woods. In the winter, the park remains open for cross-country skiing and snowshoeing along 40km of trails. *www.pinerypark.on.ca; 2hr 50min by car.*

GLAMPING ADVENTURES

There are some great places near Toronto to have a camping-like experience, without the inconvenience of pitching your own tent. At Whispering Springs Wilderness Retreat (*www. whispering springs.ca; 1hr 40min by car*), near Grafton east of Toronto, visitors stay in log-framed safari tents, equipped with private baths and electricity. There's a saltwater pool on the property, and staff deliver breakfast baskets to guests' lodgings. Located on Lake Ontario in Prince Edward County, Fronterra Farm (*www. fronterra.ca; 2hr by car*) lets guests choose between log-framed tents in the forest or floating lakeside. Showers are taken outdoors, under the stars.

15 Kingston Penitentiary

From 1835 to 2013, the maximum-security Kingston Penitentiary housed many of Canada's most notorious criminals. You can now venture behind bars on 90-minute tours of 'The Pen'. These unusual tours take you through the cell blocks, prison yards, and forbidding segregation units, as former prison guards share what it was like to work within the imposing stone walls.
www.kingstonpentour.com; 3hr
by car or by train.

16 Algonquin Provincial Park

Densely forested Algonquin Provincial Park is heaven for hikers, campers and canoe-trippers. One of Ontario's largest protected green spaces, measuring more than 7500 sq km, Algonquin has accessible trails and activities along Highway 60, which crosses the park, plus remote backcountry to explore. On select weekends, the Parkbus takes you to the park from central Toronto.
www.ontarioparks.com/park/algonquin,
www.parkbus.ca; 3hr by bus or car.

17 The Screaming Heads, Huntsville

One of Ontario's most offbeat art installations is in a field north of Huntsville, where artist Peter Camani created a series of massive concrete sculptures known as The Screaming Heads. With grimacing mouths and giant hands, the unsettling sculptures appear to be contorting their faces in pain. Visitors can wander among the sculptures but are asked not to disturb the artist, who lives in an adjacent castle. *3hr by car.*

18 Underground Railroad Historic Sites, Chatham

In the 1800s, many escaped from slavery in the United States by travelling into Canada along the Underground Railroad, a network of safe houses. You can learn about their heritage and descendants at sites like Uncle Tom's Cabin Historic Site, Buxton National Historic Site and Museum, and John Freeman Walls Historic Site and Underground Railroad Museum. *3hr by car.*

AN URBAN NATIONAL PARK?

You don't have to go far from Toronto to visit one of Canada's national parks. In fact, Rouge National Urban Park (*www.pc.gc. ca/en/pn-np/ on/rouge*) in Toronto's eastern suburbs is the country's first urban national park. More than 20 times larger than New York's Central Park, Rouge extends from the community of Markham in the north to the shores of Lake Ontario, with hiking trails through the forest and a sandy beach at the lake. Staff offer guided walks at different times of the year, too, to introduce visitors to this close-at-hand national park getaway.

16

© Mark Duffy / 500 px

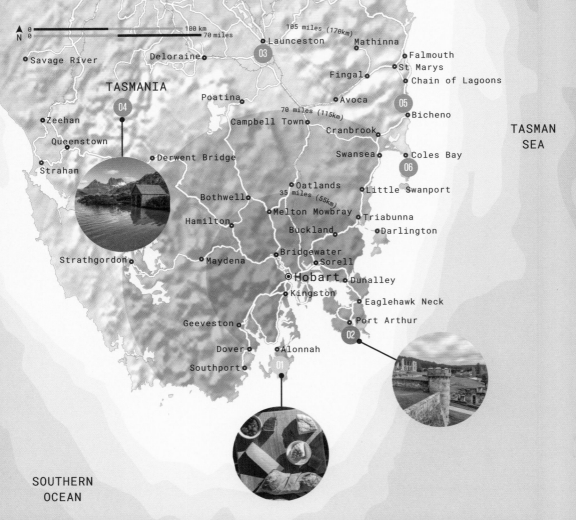

N
0 100 km
0 70 miles

105 miles (170km)

TASMANIA

TASMAN
SEA

70 miles (115km)

35 miles (55km)

Savage River
Deloraine
Launceston
Mathinna
Falmouth
St Marys
Fingal
Chain of Lagoons
Poatina
Avoca
Zeehan
Campbell Town
Bicheno
Queenstown
Cranbrook
Strahan
Derwent Bridge
Swansea
Coles Bay
Oatlands
Little Swanport
Bothwell
Melton Mowbray
Triabunna
Hamilton
Buckland
Darlington
Bridgewater
Maydena
Sorell
Hobart
Dunalley
Kingston
Eaglehawk Neck
Geeveston
Port Arthur
Dover
Alonnah
Southport
Strathgordon

01
02
03
04
05
06

SOUTHERN
OCEAN

Hobart's known for art and culture, but all around is natural Tasmanian beauty. Lose yourself deep in rainforest wilderness, sample farm-fresh produce and seafood at its source, and discover the unique history of this pocket of Australia.

HOBART

● ARTS & CULTURE   ● HISTORY   ● OUTDOORS   ● FOOD & DRINK   ● FESTIVALS & EVENTS   ● MUSIC & FILM

——— ONE HOUR FROM ———

01 Bruny Island

The home of Australia's only raw-milk cheese producer (Bruny Island Cheese Co), Bruny Island, across the D'Entrecasteaux Channel, is a spectacular introduction to Tasmania's food scene. Taste fresh oysters at the Get Shucked Oyster Farm, pick your own berries at Bruny Island Berry Farm (while eyeballing the beaches), or take your picnic to beautiful South Bruny National Park. *30min by car, then 20min car ferry from Kettering.*

02 Port Arthur Historic Site

This open-air museum on 40 hectares at the tip of the Tasman Peninsula is the best preserved convict site in Australia. Despite the grim history, the guided tours are insightful, scenic (especially boat tours around the Isle of the Dead) and entertaining, though steel your nerves for the lantern-lit ghost tour around the prison where more than 1000 died. *www.portarthur.org.au; 1hr 30min by car.*

——— TWO HOURS FROM ———

03 Cataract Gorge Reserve, Trevallyn

Not just a natural phenomenon, but home to the world's longest single-span chairlift, Cataract Gorge is awash (sorry) with activities. Brave the chilly outdoor pool in summer, take on the hiking, walking and cycling trails, admire the garden peacocks and stay until dusk for wallaby sightings. *www.launcestoncataractgorge.com.au; 2hr 20min by car.*

04 Cradle Mountain–Lake St Clair National Park

Home to Tasmania's highest peak (Mt Ossa, 1617m) and Australia's deepest lake (200m), this is the place for rugged, outdoor drama. Hike the famed Overland Track or relax on the ferry across 18km Lake St Clair. Above all, watch out for wildlife. Wombats? Yes. Echidnas? Possibly. A platypus? If you're lucky. A Tasmanian devil? No one will believe you, but maybe. *www.discovertasmania.com.au/attraction/lakestclaircradlemtlakest; 2hr 30min by car.*

05 Bicheno

Tasmania's east coast is high on quaint seaside charm. Relax on Bicehno's beaches, marvel at the Bicheno Blow Hole or watch the fairy penguins at dusk. Snorkellers are in their element here, as are experienced scuba divers, as the offshore Governor Island Marine Reserve offers some of the country's best diving. Fishers might get lucky catching fresh bream or silver trevally from Waubs Bay. *www.bichenopenguintours.com.au; 2hr 30min by car.*

——— THREE HOURS FROM ———

06 Wineglass Bay

Go on, raise a glass for a photo next to the road sign, you won't be the last. Enjoy the views as you head down through Coles Bay and Freycinet Peninsula, which takes you to an impossibly white beach ringed by pink granite mountains. The best photo (one frequently advertising Tasmania) is taken from Wineglass Bay Lookout, an hour's climb to 'the Saddle', located between Mt Atmos and Mt Mayson. *www.wineglassbay.com; 3hr by car.*

NATIONAL PARK KNOW-HOW

Remember to take all rubbish with you when you leave any of Tasmania's 19 national parks. You'll also need to buy a valid park permit to gain access. Depending on the length of your trip and the number of parks you intend to visit, there are some good-value options, such as the Holiday Pass, which covers entry into all parks for up to two months and free use of the Cradle Mountain shuttle bus. For entry fees, see https://passes.parks.tas.gov.au. Australian residents who hold a Seniors Card get 50% discount from 1 July 2019 on annual/two-year passes.

Wineglass Bay, Tasmania, three
hours from Hobart.

N

0 —————————— 150 km
0 —————————— 80 miles

Charlton

08
Bendigo

Maryborough
Avoca
Stawell
09
Castlemaine
Kyneton
Ararat
Clunes
07
02
Gisborne
Ballarat
Sunbury
Bacchus Marsh
06
Werribee
Colac
03
Geelong
05
Anglesea
Queenscliff
Lorne
04
Peterborough
10

Echuca

Shepparton

Wodonga

Wangaratta

Benalla
Myrtleford
Euroa
Bright

Mansfield
12

VICTORIA

120 miles (195km)
80 miles (130km)
40 miles (65km)

Wallan

Yarra Glen
01
Melbourne

Maffra
Sale
Warragul
Moe
Traralgon
Morwell
Korumburra
Leongatha
Yarram
Wonthaggi
Foster

11
Tidal River

BASS
STRAIT

Sandy beaches, verdant mountains, lively festivals, culinary hotspots? All are a few hours' drive of Melbourne. The city's foodie reputation includes famed wine regions on its outskirts, while pristine national parks offer wilderness adventures.

MELBOURNE

| ARTS & CULTURE | HISTORY | OUTDOORS | FOOD & DRINK | FESTIVALS & EVENTS | MUSIC & FILM |

—— ONE HOUR FROM ——

01 Coombe Yarra Valley

For wine-lovers, having the Yarra Valley less than an hour's drive from Melbourne is a boon. There are dozens of wineries to wander between here, but a standout is Coombe, the estate of the late Dame Nellie Melba, where you can taste locally produced vintages at the cellar door before proceeding into the restaurant for a marvellous high tea. ***www.coombe yarravalley.com.au; 1hr by car.***

02 Hanging Rock

Part of the picturesque Macedon Ranges, this geological monolith was created 6.5 million years ago by rapidly cooling magma. The rock is known to have been sacred to the local Wurundjeri people, who inhabited the region for some 26,000 years prior to colonisation, and recent efforts have focused on reclaiming this significant Indigenous heritage. Famously, the rock was also the setting for the Joan Lindsay novel *Picnic at Hanging Rock*. ***1hr by car.***

03 Brae

The tiny town of Birregurra, population 828, is not the most likely location for one of the world's best restaurants. And yet that's exactly where you'll find Brae, the multi-awarded brainchild of chef Dan Hunter. Ever-changing degustation menus highlight seasonal and unusual ingredients, grown on the property or by local farmers. Extend the experience by spending a night in one of the beautiful suites overlooking the surrounding hills. ***www.braerestaurant.com; 1hr 30min by car.***

04 Peninsula Hot Springs

On a drizzly Melbourne day, when the air is too fresh for ocean swimming, there's nothing better than a soak at these geothermal springs on the Mornington Peninsula. Inspired by Japanese onsen, a series of pools is scattered across the landscape, allowing you to wander between baths of varying temperatures. For further pampering, spa treatments are also available. ***www.peninsulahotsprings.com; 1hr 30min by car.***

05 Blues Train, Queenscliff

Climb aboard the Blues Train at the historic Queenscliff railway station for a memorable ride along the restored Bellarine Railway. During the evening, passengers rotate through a series of four carriages, each playing host to a different musical act. Dance the night away as the train travels to Drysdale and back. On your return, the charming main street of Queenscliff is just a stumble away. ***www.thebluestrain.com.au; 1hr 30min by car.***

06 Meredith Music Festival

More than just a festival, Meredith is an immersive experience. Spend three days camping in the Victoria bush surrounded by live performers and beautiful people. The festival has been running since 1991 and tickets are distributed via a highly contested ballot; if you miss out, try for Golden Plains, in early March at the same site. ***www.mmf.com.au; early Dec; 1hr 30min by car.***

07 Lake House, Daylesford

Daylesford has long held a reputation for relaxation, thanks in no small part to the natural mineral springs that are dotted all

Courtesy of Brae; © FiledIMAGE / Shutterstock; Courtesy of The Blues Train over the surrounding area. The town's pièce de résistance is the award-winning Lake House restaurant, helmed by chef Alla Wolf-Tasker, where multi-course degustation menus are served with finesse and flair. Stay overnight in one of the attached suites to experience the ultimate luxury break. *www.lakehouse.com.au; 1hr 30min by car.*

——— TWO HOURS FROM ———

08 **Bendigo Art Gallery**
Established in 1887, Bendigo Art Gallery is one of Australia's oldest regional art galleries. An $8.5 million redevelopment in 2014, which added 600 sq m of exhibition space, means it is also among the largest. A varied permanent collection of more than 5000 works is bolstered by world-class temporary exhibitions, which rotate regularly through its light-filled rooms. *www.bendigoartgallery.com.au; 1hr 50min by car or 2hr from Southern Cross Station on V-line.*

09 **Clunes Booktown**
More than 18,000 bibliophiles converge on the diminutive town of Clunes every autumn for the Booktown Festival, which features author talks, panel discussions, displays of rare and collectible books, and many more bookish activities besides. Clunes actually draws literary types all year round, who come to peruse its disproportionately high number of bookshops and attend the regular 'Booktown on Sundays' events. *www.clunesbooktown.com.au; late Apr or early May; 1hr 50min by car.*

——— THREE HOURS FROM ———

10 **Twelve Apostles**
These craggy limestone formations jut from the water alongside the Great Ocean Road, which meanders some 250km between Torquay and Warrnambool. There are only eight apostles nowadays, but the scenery is as incredible as ever. It's worth overnighting in nearby Port Campbell to visit early in the day and avoid the tour buses. *3hr by car.*

© Chris Williams Black Box / Getty Images

11 Wilsons Promontory National Park
Melburnians are spoilt for choice when it comes to beach getaways, but to really escape the urban sprawl you can't do better than the Prom, where long stretches of white sand await. Even those who prefer some creature comforts can experience the delights of the park at Tidal River, where cabins tucked among coastal bushland give a sense of peaceful seclusion.
www.parkweb.vic.gov.au; 3hr by car.

12 Mt Buller
Melbourne's winters can be grey and dreary, but only a few hours away is a veritable winter wonderland of white, bright cold stuff. Mt Buller's ski fields are some of the best in the country, offering 300 hectares of skiable terrain and 22 lifts to hoist you up the mountainside. There are also plenty of après-ski options to keep you occupied once the sun goes down.
www.mtbuller.com.au; 3hr by car.

SURFING GETAWAY

Sydney gets all the publicity when it comes to beaches, but Victoria's coastline offers plenty of opportunities for experienced and novice surfers alike. For those with surfing chops, Bells Beach (1hr 30min by car), near Torquay, has a consistent right-hand break and is home to the world's longest continuously running surf tournament. The long, shallow main beach at nearby Anglesea (1hr 40min by car), meanwhile, is renowned for its beginners' waves – hence the surf schools nearby. Once you've mastered the basics, a series of beautiful beaches all the way to Apollo Bay (3hrs by car) offers the chance to practise your skills.

11

© Natphotos / Getty Images

AUSTRALIA'S BEST FOOD & DRINK ESCAPES

It might be at the bottom of the world but an encounter with Australia's gourmet scene will soon push Oz to the top of your must-return list.

Pt Leo Estate, Merricks, Victoria

Sublime setting? Winery? Sculpture park? Yes, yes and yes. There's also Laura, the fine-dining restaurant that has topped food critics' lists since its opening in February 2018. Two other casual eating options offer views across the Mornington Peninsula coastline. Spectacular. *www.ptleoestate.com. au; sculpture park $10/ free for diners 11am-6pm daily; Laura lunch & dinner Thu-Sat, lunch Sun; 1hr from Melbourne by car.*

Royal Mail Hotel, Dunkeld, Victoria

The home of Australia's largest wine cellar (28,000 bottles), the Royal Mail hosts gourmands keen for a country getaway in the Grampian National Park. Choose from the upmarket Wickens at Royal Mail Hotel or the more casual Parker Street Project. *www.royalmail.com. au; Wickens dinner Wed-Sat, lunch Sat; Parker Street Project breakfast, lunch & dinner daily; 3hr from Melbourne by car.*

The Long Apron, Montville, Queensland

Head to Queensland's Sunshine Coast for innovative dining at The Long Apron, where you might start your meal with a 'tree' in a vase (actually caraway grissini) and end with rainforest 'twigs' that turn out to be white chocolate. *www.spicersretreats. com/restaurants/the-long-apron; breakfast & dinner Mon-Sun, lunch Fri-Sun; 1hr 30min from Brisbane by car.*

Margaret River Gourmet Escape, Western Australia

If it's good enough for Nigella...Yes, she's a regular guest, along with many top Australian chefs, at this West Coast food festival held over three days in November. The Margaret River region also produces some of the country's best wines. *www. gourmetescape.com. au; Nov; 3hr from Perth by car.*

Fleet, Brunswick Heads, New South Wales

Don't judge the tiny space; this 22-seater (with space for eight outside) has been wowing diners since it opened in 2015. If you can cope with one single communal table that leads from the wine bar into the kitchen, sophisticated cooking awaits. *www.fleet-restaurant. com.au; late lunch (from 3pm) & dinner Thu-Sat, lunch (from 12pm) Sun; 20min from Byron Bay by car.*

Provenance, Beechworth, Victoria

The charming Victorian country town of Beechworth is a surprising location for an award-winning restaurant infusing native Australian ingredients with Japanese flavours and techniques. Chef Michael Ryan's creations are inspired by his frequent trips to Japan, where he often leads food tours. *www.theprovenance. com.au; dinner Wed-Sat, lunch Sun; 3hr from Melbourne by car.*

Agrarian Kitchen Cooking School & Farm, Lachlan, Tasmania

Regularly topping Tasmania's 'best dining' lists (amid stiff competition), Agrarian Kitchen is a five-acre working farm where the paddock-to-plate ethos is strong. Here you'll cook with heirloom veg and rare breeds of meat. There's a sensational sister restaurant nearby in New Norfolk. *www.theagrariankitch-en.com; 50min from Hobart by car.*

Paper Daisy, Cabarita Beach, New South Wales

Beachfront Halcyon House is an idyllic retreat on NSW's North Coast and its restaurant Paper Daisy has been absolutely raking in the national accolades since its inception in 2015. Try the paperbark grilled fish for a true sense of place. *www.halcyonhouse. com.au/paper-dai-sy-restaurant; break-fast, lunch & dinner daily; 40min from Byron Bay by car.*

Stone House Wine Bar & Kitchen, Darwin, Northern Territory

Melburnian Rebecca Bullen's idea to launch Darwin's first proper wine bar has changed the Top End's drinking scene: it's a staggeringly good find and has been winning wine list awards ever since she launched it in September 2016. *www. stonehousedarwin. com.au; 4pm-late Mon-Thu & Sat-Sun, from 3pm Fri.*

Maggie Beer's Farm Shop, Nuriootpa, South Australia

Australian celebrity chef and cookbook writer Maggie Beer has a farm shop in the beautiful Barossa, with an adjoining restaurant and studio (for daily demonstra-tions of her verjuice and vino cotto). It's tastings paradise. The Orchard House accommodation is a stroll away through the pear trees. *www.maggiebeer.com. au; 10.30am-5pm; 1hr from Adelaide by car.*

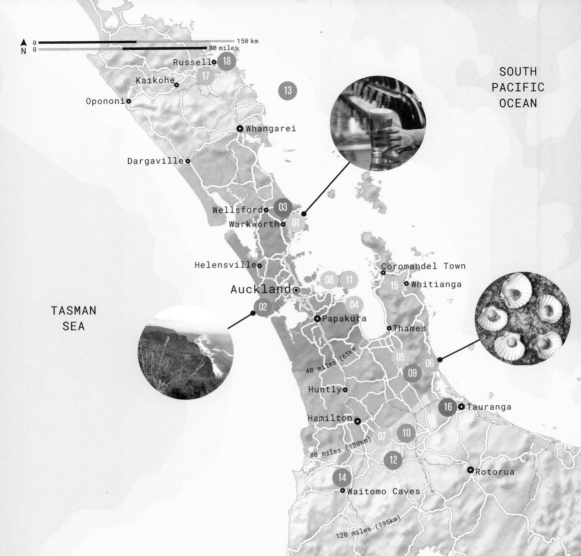

New Zealand's most cosmopolitan city is a great base from which to discover the colonial and Maori history of the Northland, some fabulous eating and drinking, and exciting outdoor adventures — including underground and underwater.

AUCKLAND

Map labels:

TASMAN SEA

SOUTH PACIFIC OCEAN

N

0 — 150 km
0 — 80 miles

Opononi
Kaikohe
Russell · 17 · 18
13
Whangarei
Dargaville
Wellsford · 03
Warkworth · 01
Helensville
Coromandel Town
Auckland · 08 · 11
Whitianga · 15
02
04
Papakura · Thames
05
40 miles (65km) · 09 · 06
Huntly
16 · Tauranga
Hamilton
07 · 10
80 miles (130km)
12
Rotorua
14
Waitomo Caves
120 miles (195km)

● ARTS & CULTURE   ● HISTORY   ● OUTDOORS   ● FOOD & DRINK   ● FESTIVALS & EVENTS   ● MUSIC & FILM

——— ONE HOUR FROM ———

01 Sawmill Brewery
Defer a visit to the vineyards and farmers' markets around rural Matakana by ordering a tasting tray of craft beers from the Sawmill Brewery. Seasonal releases complement year-round brews, such as the hearty Weizenbock or refreshing Pilsner, and the brewery's rustic Smoko Room restaurant turns out surprising food packed with international flavours. Look for the hop vines out front and you're in the right place. *www.sawmillbrewery.co.nz; 1hr by car.*

02 Hillary Trail
Named after the late Sir Edmund Hillary, one of the first people to conquer Mt Everest, the Hillary Trail offers superb bush and coastal views just a short drive from New Zealand's biggest city. Extending 77km in total, the trail usually takes from four to six days, but shorter day-long walks can also be enjoyed. *www.piha.co.nz/hillary-trail; 1hr by car.*

03 Sculptureum
Get ready to be surprised at the Sculptureum, a labour of love for its owners – two art aficionado lawyers. Six galleries, where artists represented include Picasso and Chagall, combine with three manicured gardens filled with challenging installations and thought-provoking sculptures from around the world. After a few hours' contemplating the artistic and intellectual highlights, satisfy your inner gourmand at the Sculptureum's excellent Rothko restaurant. *www.sculptureum.nz; Thu–Mon; 1hr by car.*

04 Splore
Held annually under the crimson blooms of pohutukawa trees, Splore fills the shaded cove of Tapapakanga Regional Park with a chilled programme of live music, DJs and visual arts. The line-up is always a cosmopolitan mix, combining headlining global acts with the very best of local talent, and for the three days it's held in late February, Splore is quite probably the most laid-back locale in all the land. *www.splore.net; Feb; 1hr 20min by car.*

——— TWO HOURS FROM ———

05 The Refinery
Retro 1970s furniture and Kiwiana decor both feature at this brilliant cafe in a country town known for its vintage clothing stores and antique shops. Fire up the turntable with your pick from hundreds of vinyl records, and partner a robust espresso with a grilled Cuban sandwich. During spring and summer, the pleasantly overgrown garden is definitely the place to be. *www.the-refinery.co.nz; 1hr 40min by car.*

06 Waihi Gold Discovery Centre
Since 1878, the lustre of silver and gold has illuminated the mining economy of Waihi, and the Gold Discovery Centre is a seriously entertaining and informative showcase of the region's gold-flecked history. After negotiating the centre's various interactive displays – including innovative holograms and the opportunity to beat a 'virtual miner' at gambling – join a tour to explore the local gold mining scene. *www.golddiscoverycentre.co.nz; 1hr 40min by car.*

THE LOCAL'S VIEW

'For me, New Zealand's Karekare Beach is the most beautiful beach in the world. Just an hour's drive from Auckland, this idyllic spot overlooking the Tasman Sea is tucked away in the Waitakere Ranges. The coastline's rugged rocks and black sand make it the perfect getaway from our buzzing city centre. Nearly every year for two decades, the Karekare Beach Races have been a spectacle not to miss. There's nothing like galloping at full speed with the wind in your hair and the smell of the sea filling your every breath. Priceless.'

Giapo Grazioli, Giapo Ice Cream

© Neil O'Shea / Alamy Stock Photo; Courtesy of Sawmill Brewery; © Katerinina / Shutterstock

07 Cambridge Farmers Market

Celebrating an English heritage vibe, pretty Cambridge also hosts one of New Zealand's best farmers' markets. Get there early for a relaxed Saturday morning breakfast, stopping in at the various stalls gathered on Cambridge's leafy Victoria Square. You'll find the best coffee in town at the Manuka Brothers' stall, and the savoury treats from the Raglan Pie Co are deservedly famous around these parts. *www.waikatofarmersmarkets.co.nz; 8am–noon Sat; 2hr by car.*

08 Tantalus Estate Vineyard

On an island bursting with excellent vineyards and great eating, Tantalus Estate stands out courtesy of its flavour-filled menu, surprising architectural design and its own in-house craft brewery. The highlights of the wine list are Rhône- and Bordeaux-style red varietals. Beers made by Alibi Brewing partner well with seasonal dishes imbued with global influences – you won't regret trying the crisp IPA with the pulled jackfruit tacos. *www.tantalus.co.nz; 2hr by ferry & car.*

09 Karangahake Gorge

Framed by native forest and soundtracked by the Karangahake river, one of New Zealand's best scenic drives is also perfect for bushwalking or mountain biking. To get there, your best bet is to catch the vintage train from nearby Waihi to Waikino Station, where bikes can then be hired. Follow the trail fringing the river and finish with wood-fired pizza and local beers at the Bistro at the Falls Retreat. *www.doc.govt.nz; 2hr by car.*

10 Hobbiton

A magnet for Middle Earth fans from across the planet, the Hobbiton film set near the Waikato farming town of Matamata is a charming slice of moviemaking magic. Guided tours packed with inside knowledge illuminate the stories behind the cinematic versions of JRR Tolkien's books. Don't miss downing a brew at the Green Dragon Inn, or signing up for one of Hobbiton's popular evening banquets. *www.hobbitontours.com; 2hr by car.*

© Danita Delimont / Getty Images

09

11 Waiheke Wine & Food Festival
Auckland's island of art, wine and food is enlivened in late summer with this annual festival, usually held across five days. Beyond the sybaritic focus of events such as Italian-themed long lunches, wine-blending workshops and multi-course dinners, other cultural elements include jazz concerts, coastal sculpture installations and vintage markets. Adrenaline junkies should sign up for Waiheke's zipline to enjoy vineyard and island views. *www.facebook.com/waihekewineand foodfestival; Mar; 2hr by ferry & car.*

12 Sanctuary Mountain
Protected by 47km of pest-proof fencing, the forested triple peaks of Maungatautari comprise one of New Zealand's most beloved natural sanctuaries. Crossing the mountain on well-maintained tracks takes about six hours; or join one of the day walks exploring the bird and insect life around the sanctuary's Southern Enclosure. There's also a 'tuatarium' where the country's unique reptile, the tuatara, can be seen. *www.sanctuarymountain.co.nz; 2hr 30min by car.*

ISLAND LIFE AROUND AUCKLAND

Auckland is the departure point for getaways to the islands of the Hauraki Gulf. Car and passenger ferries venture to Waiheke Island (2hr by ferry and car) for beaches, galleries and vineyards, while Rangitoto's volcanic cone is best reached on a kayaking trip. To the north, forested Kawau Island (30min by ferry) combines colonial history with interesting walks, while various avian species, including the endangered takahe, are the stars on Tiritiri Matangi (1hr 20min by ferry). On Auckland's eastern horizon, Great Barrier (30min by plane) has the world's only island Dark Sky Sanctuary.

13 Poor Knights Islands Marine Reserve

Rated as one of the world's top 10 diving spots, the islands' location, amid a subtropical current surging from the Coral Sea, produces a more diverse range of underwater life than in other New Zealand waters. Undersea archways, tunnels and caverns all shelter a surprising array of marine life. Operators based in nearby Tutukaka include kayaking, paddleboarding and snorkelling among their Poor Knights adventures. **www.diving.co.nz/ poor-knights-islands; 2hr 40min by car.**

14 Waitomo Caves

Studded with underground caves and subterranean rivers, New Zealand's Waitomo region is a superb location for intrepid travellers. Exciting options to explore the area's underground labyrinth include rafting on inner tubes through hidden rivers, abseiling into forested limestone caverns, and even a thrilling underground flying fox (zipline). Leisurely strolls and silent boat trips through Waitomo's fascinating glow-worm caves provide alternative quintessential New Zealand experiences. **www.waitomo.com; 2hr 40min by car.**

15 Whitianga Scallop Festival

Get ready for shellfish shenanigans as one of the world's favourite bivalves is the main focus at this culinary extravaganza on New Zealand's Coromandel Peninsula. Fresh Whitianga scallops are given a menu of tasty makeovers and partnered with local beer and wine. The one-day festival also

© Beat J Korner / Shutterstock

© Shaun Jeffers / Shutterstock

13

14

includes plenty of live music and entertainment. Don't leave town without trying a barbecued scallop and bacon kebab. *www.scallopfestival.nz; Sep; 2hr 40min by car.*

16 Tauranga Art Gallery
In a diverse coastal region best known for surf beaches and volcanic activity, one of New Zealand's best regional art galleries provides cultural and cerebral balance. Tauranga does have a slightly conservative reputation but the city's gallery presents an ongoing programme of challenging and innovative contemporary exhibitions. Adjourn to the terrific cafes and restaurants along The Strand to mull over what you've just seen. *www.artgallery.org.nz; 2hr 50min by car.*

17 Te Kōngahu Museum of Waitangi
Signed in 1840, New Zealand's Treaty of Waitangi is regarded as the country's still-relevant founding document, and this modern museum, opened in 2016, is a comprehensive showcase of the role of the treaty in the past, present and future of New Zealand. Many historical *taonga* (treasures) from across the country are now collected here. The surrounding Treaty Grounds provide views of the Bay of Islands. *www.waitangi.org.nz; 3hr 10min by car.*

18 Russell Nature Walks
Exploring with Russell Nature Walks is an opportunity to get up close and personal to native New Zealand birds, including the weka and the tui. Book in for a night tour, and with the diffuse illumination of glow-worms, there's the chance to hear NZ's national bird, the kiwi. The country's flightless, feathered icon is notoriously shy though, so actually seeing one in the wild is a rare occurrence. *www.russellnaturewalks.co.nz; 3hr 30min by car.*

AVOIDING THE TRAFFIC

Sprawling north and south across a narrow isthmus, Auckland is one of the world's most impressive harbour cities, but this spectacular geography can create challenges when escaping the city by car. Heading north, try to avoid holiday weekends, especially during summer, and schedule trips to popular Waiheke Island on a weekday. Note that some Waiheke vineyard restaurants are only open from Wednesday to Sunday outside summer (November to March). In midsummer, the narrow roads of the Coromandel Peninsula groan with caravans and campervans, so consider visiting in December or March.

18

© tiire / Shutterstock

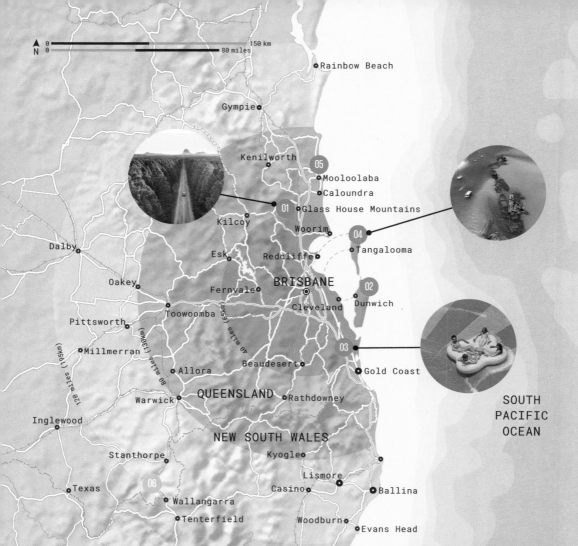

N

0		150 km
0		80 miles

Rainbow Beach

Gympie

Kenilworth

05

Mooloolaba

Caloundra

01 Glass House Mountains

Kilcoy

Woorim

04

Esk

Redcliffe

Tangalooma

Dalby

BRISBANE

Oakey

Fernvale

Cleveland

02

Pittsworth

Toowoomba

Dunwich

Millmerran

03

Allora

Beaudesert

Gold Coast

Warwick

QUEENSLAND

Rathdowney

SOUTH
PACIFIC
OCEAN

NEW SOUTH WALES

Stanthorpe

Kyogle

06

Lismore

Texas

Casino

Ballina

Wallangarra

Tenterfield

Woodburn

Evans Head

Inglewood

Even party people need to escape energetic, young and full-of-fun Brisbane sometimes, and Queensland offers plenty of choice: granite winelands, pristine national parks, sandbar islands and the beaches of the Gold and Sunshine Coasts.

BRISBANE

● ARTS & CULTURE ● HISTORY ● OUTDOORS ● FOOD & DRINK ● FESTIVALS & EVENTS ● MUSIC & FILM

─── ONE HOUR FROM ───

01 **Glass House Mountains National Park**

Brisbane is encircled by national parks offering serene silence and eucalyptus scents, but the Glass House Mountains add something otherwordly. This garden of volcanic plugs bursts up from the plain, with walking trails and climbing routes ascending to the summits of several dramatic peaks. Beerwah is best for first-timers, despite some nerve-jangling exposed sections on the trail. *www.npsr.qld.gov.au/parks/ glass-house-mountains; 1hr 30min by car.*

─── TWO HOURS FROM ───

02 **North Stradbroke Island**

'Straddie' is a favourite weekend getaway for Brisbane locals, passed over by the outsiders rushing straight to Noosa and Fraser Island. There's foodie fun around Point Lookout, but the best of the island is further south, with swimming lakes and 4WD tracks flanking the endless sandy beach on the eastern shore. *www.stradbrokeisland.com; 2hr by train, bus & ferry from Roma St or Central station.*

03 **Dreamworld, Coomera**

You won't escape the crowds at Dreamworld, Queensland's biggest theme park, but you'll definitely dodge the humdrum on stomach-flipping rides like the Giant Drop and Tower of Terror. With 13 zones overflowing with adrenaline-rush experiences, you're sure to find something to thrill/nauseate. *www. dreamworld.com.au; 10am–5pm; 2hr by train & bus from Brisbane Central Station.*

04 **Tangalooma, Moreton Island**

The name 'Moreton Bay' is spoken in hushed tones by Brisbane foodies thanks to the Moreton Bay Bug – a sublimely delicious species of slipper lobster – but there's more to this strip of coast than delectable crustaceans. Moreton Island, the tranquil sandbar guarding the entrance to the bay, features empty beaches, visiting dolphins, calm campgrounds and snorkelling on a string of rusting shipwrecks. *2hr by boat from Port of Brisbane.*

05 **Mooloolaba, Sunshine Coast**

Whereas the Gold Coast calls out to the brash and the beautiful, the Sunshine Coast is where ordinary folks come to unwind. Mooloolaba is the definitive Sunshine Coast beach: blonde sand, clean waters, easy-going locals, a cafe-filled esplanade and a load of surf schools to keep the kiddies on board. The sun 'n' surf action spills right around the promontory to Alexandra Headland and Maroochydore. *www.visitsunshinecoast. com/mooloolaba; 2hr by bus.*

─── THREE HOURS FROM ───

06 **Ballandean Estate, Ballandean**

Queensland wine sits in the shadow of famous quaffs from Margaret River and the Hunter and Barossa valleys, but locals swear by the wines produced from 'strange birds' (little-known, alternative grape varieties) that flourish in the rocky highlands of Stanthorpe's Granite Belt. Ballandean has been run by the same Italian family since the 1930s, and its Viognier is an ideal start to a Granite Belt wine tour. *www.ballandean estate.com; 9am–5pm; 3hr by car.*

BRISBANE BY RAIL & RIVER

The Brisbane River is the city's reason for existing, providing access 344km inland through the foothills of the Great Dividing Range. But navigating it was no easy task for Queensland's early settlers; prone to silting and flooding, the waterway was also notorious for bull sharks, still found as far as 32km inland today. To explore modern Brisbane, jump aboard CityCat ferries, which travel the river between St Lucia and Northshore Hamilton, serving the inner suburbs. CityTrains run to the Sunshine Coast and Gold Coast, also giving access to the jetties for ferries to Moreton Bay.

© FLYFILM.TV / Getty Images; © Darren Tierney / Shutterstock; Courtesy of Dreamworld

OCEANIA'S MOST FASCINATING INDIGE

Explore the diverse countries of the vast Pacific Ocean to discover the region's authentic and vibrant indigenous cultures.

● ARTS & CULTURE   ● HISTORY   ● OUTDOORS   ● FOOD & DRINK   ● FESTIVALS & EVENTS   ● MUSIC & FILM

——— ONE HOUR FROM ———

01 Feral Brewing Company
Producing brews dubbed Watermelon Warhead and Barrique O'Karma, Feral Brewing Company is one of Western Australia's most innovative craft breweries. The Hop Hog Pale Ale is a modern Australian classic, enjoyed by loyal locals who frequent Feral's rustic Swan Valley location most weekends. Nearby you'll find purveyors of artisan cheese and chocolate, so make a day of it.
www.feralbrewing.com.au; 1hr by car.

02 Rottnest Island
Catch a fast ferry from either Perth or nearby Fremantle to explore Rottnest Island. A bus circles the island's main attractions, including superb beaches and surveys of Rottnest's historical legacy of salt production, but independent discovery by mountain bike is also popular. Bikes can be hired on the island or from ferry operators. Don't miss saying g'day to the quokkas, a species of marsupial found almost exclusively on Rottnest. *1hr 30min by ferry.*

——— TWO HOURS FROM ———

03 New Norcia
New Norcia was established as a missionary settlement for the local Aboriginal community by Spanish Benedictine monks in 1846 – visit the mission's museum, art gallery and church by guided tour. You can stay at the historic New Norcia Hotel; an essential dining experience is pairing New Norcia's wood-fired bread with the Belgian-style Abbey Ale brewed exclusively for the hotel. *www. newnorcia.wa.edu.au; 1hr 50min by car.*

——— THREE HOURS FROM ———

04 Dryandra Woodland
The Barna Mia Animal Sanctuary in the Dryandra Woodland area affords a chance to see endangered Aussie wildlife such as bilbies, boodies and woylies – names straight out of *Dr Dolittle*. Add in a few threatened numbats, and the area's a must for fans of quirky marsupials. Accommodation includes renovated woodcutters' cabins in a 1920s forestry camp. Book an after-dark tour for the best action. *https://parks.dpaw.wa.gov.au/park/ dryandra-woodland; 2hr 10min.*

05 Nambung National Park
In a state with no shortage of natural highlights, the landscapes of Nambung's Pinnacles Desert are still truly spectacular. Against a background of a cobalt Indian Ocean, thousands of wind-eroded limestone pillars stand sentinel to time, and splashes of colour on this largely monochrome moonscape are provided by striking grey and pink parrots. Visit at dawn or sunset for the best experience and smaller crowds. *https://parks.dpaw.wa.gov.au/park/ nambung; 2hr 30min by car.*

06 Nannup Music Festival
A pretty riverside town, Nannup's reputation as a top spot for bushwalking and canoeing is usurped by this autumn festival showcasing both Australian folk and world music. It's the kind of gig where busking and an independent spirit is encouraged – so if you're feeling brave, the stage might be yours for the taking. *www. annupmusicfestival.org; Mar; 3hr by car.*

ACTIVE AROUND MARGARET RIVER

The Margaret River region (3hr by car) offers a sybaritic blend of wine, craft beer and artisan food, but Perth locals' favourite culinary getaway offers adventure too. Running from Cape Naturaliste south for 135km next to the Indian Ocean, the Cape to Cape Track can be broken into day hikes. There's mountain biking in the Boranup Forest, while kayaking, abseiling and rock climbing all negotiate the craggy coastline. Underground, the CaveWorks complex (www. margaretriver. com/attractions/ caves) includes colourful Lake Cave and the 86m-deep Giants Cave, accessible via vertical ladder climbs.

© Catherine Sutherland, © Bob Christopher / 500 px, © Pinkcandy / Shutterstock

SOUTH→
AMERICA↓

NOUS EXPERIENCES

Footprints Waipoua, New Zealand

Tribal heritage and Māori spirituality underpin these twilight cultural tours exploring one of New Zealand's most spectacular native forests. Led by local Māori guides, tour highlights include heartfelt karakia (prayers) before NZ's biggest trees. *http://footprint-swaipoua.co.nz; 6pm Oct-Mar, 5pm Apr-Oct; 1hr 10min by car from Paihia.*

Toi Hauāuru Arts Studio, New Zealand

Near the NZ surf town of Raglan, artist Simon Te Wheroro crafts contemporary Māori artwork and sculpture, and is also a skilled practitioner in the art of tā moko (tattooing). *www.facebook.com/ ToiHauauruStudio; 10am-5pm Wed-Sun; 1hr by car from Hamilton.*

Mitai Māori Village, New Zealand

One of Rotorua's most friendly and talented Māori families combines cultural performances, a traditional hangi (Māori feast) and even a glow-worm bushwalk during their nightly three-hour shows. *www.mitai.co.nz; 6.30pm; 30min by car from Rotorua.*

Highland Paradise Cultural Centre, Cook Islands

Astounding hilltop views over Rarotonga's reef-encircled coastline combine with Cook Islands' cultural performances and a traditional *umukai* (underground earth oven) feast at the Highland Paradise centre. *www.facebook.com/ HighlandParadise; 5.30-9.30pm Mon, Wed & Fri; 30min by car from Avarua.*

Sigatoka River Safari, Fiji

Take a 45km spin by jetboat up Fiji's Sigatoka river to a local village. Different settlements are visited throughout the week, and village chiefs welcome guests with a kava (a mildly narcotic Pacific drink) session. *www.sigatokariver. com; 8.45am-2pm; 1hr by car from Nadi.*

Ekasup Cultural Village, Vanuatu

One of Vanuatu's best cultural experiences incorporates visits to this *kastom* (traditional) village. Ekasup's population trace their ancestry back to the island of Futuna in remote central Polynesia. *www.tourismvanuatu. com; 9am-noon Mon-Fri; 10min by car from Port Vila.*

Jean-Marie Tjibaou Cultural Centre, New Caledonia

A tribute to the pro-independence leader assassinated in 1989, this centre was designed by Italian architect Renzo Piano, and is the Pacific's most impressive showcase of New Caledonia's indigenous Kanak culture. *www.dck.nc; 9am-5pm Tue-Sun; 10min by car from Noumea.*

Koomal Dreaming, Australia

Local man Josh Whiteland, from the Wadandi tribe around Western Australia's Margaret River region, runs interesting tours combining Aboriginal food, culture and music. Bushwalking and exploring the nearby Ngilgi Cave are also on offer. *www.koomaldreaming. com.au; 1hr by car from Bunbury.*

Guringai Aboriginal Tours, Australia

Exploring the Ku-ring-gai Chase National Park north of Sydney, these tours incorporate visits to ancient rock art sites, and explain how Australia's indigenous Aboriginal people are careful custodians of their natural environment. *www.guringaitours. com.au; 1hr by car from Sydney.*

Brambuk National Park & Cultural Centre, Australia

Run by the indigenous Koori people of Hall's Gap, Victoria, this contemporary museum, gallery and cultural centre provides plenty of hands-on activities, including painting, boomerang-throwing and playing the didgeridoo. *www.brambuk.com.au; 9am-5pm; 3hr by car from Melbourne.*

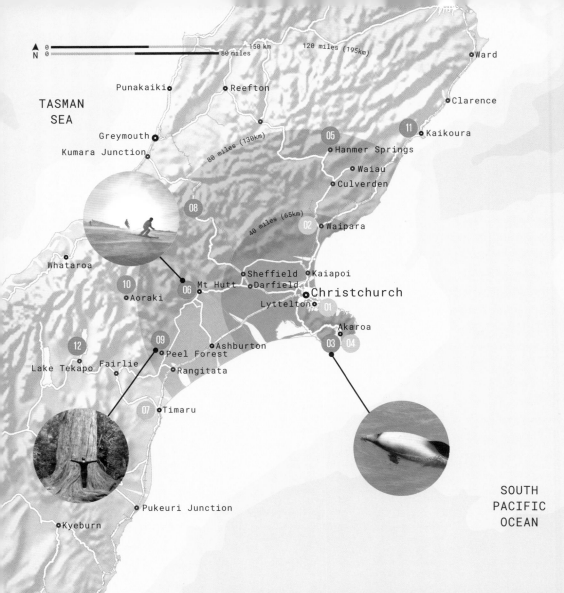

TASMAN SEA

N

0 ——— 150 km
0 ——— 80 miles

120 miles (195km)

Ward

Punakaiki • Reefton •

Clarence •

Greymouth •

11 Kaikoura •

Kumara Junction •

80 miles (130km)

05 Hanmer Springs •

Waiau •
Culverden •

Whataroa •

08

06 Mt Hutt
Sheffield • Kaiapoi •

02 Waipara •

40 miles (65km)

10 Aoraki •

Darfield •

Christchurch

Lyttelton •

01

Akaroa •

03 04

12 Lake Tekapo

09 Peel Forest •

Ashburton •

Fairlie •
Rangitata •

07 Timaru •

SOUTH
PACIFIC
OCEAN

Pukeuri Junction •

Kyeburn •

Right at the heart of the South Island, Christchurch has the glittering Pacific Ocean on one side and the peaks of the Southern Alps on the other. You'll find spectacular scenery, charming towns and abundant wildlife whichever way you turn.

CHRISTCHURCH

● ARTS & CULTURE   ● HISTORY   ● OUTDOORS   ● FOOD & DRINK   ● FESTIVALS & EVENTS   ● MUSIC & FILM

ONE HOUR FROM

01 Roots

Just over Christchurch's Port Hills, the harbour suburb of Lyttleton can feel a world away. The friendly main street has a plethora of charms, but the jewel in its crown is Roots, the award-winning restaurant of chef Giulio Sturla and his wife Christy. Their degustation menus alone are worth the trip over the hills. *www.rootsrestaurant.co.nz; 20min by car or 30min by bus.*

02 Pegasus Bay

The verdant Waipara Valley is renowned for cool-climate wines, such as Riesling and Pinot Noir. It has more than 20 different wineries scattered across its rolling hills, so it makes the perfect weekend away for wine lovers. Pegasus Bay, where the Donaldson family have been tending their vines since 1986, is the obvious choice for lunch (Thursdays to Mondays only); its light-filled restaurant set among manicured gardens is the ultimate treat. *www.pegasusbay.com; 1hr by car.*

03 Akaroa Dolphins

The picturesque Banks Peninsula is a nature-lover's paradise, with a whole host of wild animals calling the peninsula's coves and harbours home. Take to the waters on a cruise with Akaroa Dolphins and you'll hopefully encounter the endangered Hector's dolphin, the world's smallest and rarest, in addition to New Zealand fur seals, white-flippered penguins and the most important crew member of all: a life-jacket-wearing, dolphin-spotting dog. *www.akaroadolphins.co.nz; 1hr 20min by car.*

04 Akaroa French Fest

Akaroa's French heritage (it was the site of New Zealand's only French colony) comes right to the fore during this celebration of all things Gallic: music, markets, activities, crafts, theatre and – *bien sûr* – food and wine. The festival is held biennially during odd-numbered years, but even at other times Akaroa exudes a distinctly French village air, right down to its francophone street names. *www.frenchfest.co.nz; mid-Oct in odd-numbered years; 1hr 20min by car.*

TWO HOURS FROM

05 Hanmer Springs Thermal Pools

Māori legend says that these bubbling hot springs are the result of burning embers from Mt Ngauruhoe on the North Island falling to earth. The waters certainly are hot – up to 42°C – and are believed to have numerous therapeutic properties. Once you have soaked all your worries away, the pretty spa town of Hanmer Springs has plenty to divert you further. *www.hanmersprings.co.nz; 1hr 50min by car or 2hr by bus.*

06 Mt Hutt

The largest commercial ski field on the South Island, offering 365 skiable hectares of terrain, Mt Hutt is the place to be when the snowflakes begin to fall – its slopes are the first in the southern hemisphere to open each year. The nearby town of Methven comes alive during the ski season, when its bars and pubs are packed to the rafters with snow bunnies making the most of the lively après-ski scene. *www.mthutt.co.nz; 2hr by car.*

SCENIC RAIL JOURNEYS

Two of the land's most scenic train journeys (*www.greatjourneyofnz.co.nz*) depart Christchurch. The first is the TranzAlpine, which travels west through a series of tunnels and gradients to Greymouth. The five-hour journey takes in increasingly dramatic landscapes, from the verdant Canterbury Plains to the forested Southern Alps. Or you could take the five-hour Coastal Pacific journey tracing the coastline north to Picton, where it intersects with the Interislander Ferry. With sea on one side and lush, mountainous terrain on the other, you'll hardly know which way to look.

© kovop58 / Shutterstock; © Carefordolphins; © Carefordolphins / Alamy Stock Photo; © Happy Auer / Shutterstock

07 Te Ana Māori Rock Art Centre
The area surrounding Timaru is home to a wealth of Māori rock art sites, which are brought to life at this immersive multimedia gallery and museum. Passionate Ngāi Tahu guides retrace the journeys of their ancestors through stories and images; visitors also have the opportunity to participate in extended self-drive tours, which visit nearby rock art galleries in situ. *www.teana.co.nz; 2hr by car or 2hr 30min by bus.*

08 Arthur's Pass National Park
Arthur's Pass – at 900m, New Zealand's highest settlement – is the perfect jumping-off point for popular day walks and multi-day tramps over varied alpine valleys and ridges. It's also a great place to spot kea (alpine parrots) – keep an eye on these cheeky fellas near your gear, lest they decide to use their sharp beaks to investigate further. *www.doc.govt.nz; 2hr by car or 2hr 30min by bus.*

09 Peel Forest
Keen hikers, twitchers and outdoorsy types will love Peel Forest, a small but significant section of indigenous podocarp forest that is home to ancient trees, some hundreds of years old. Short day treks to several nearby waterfalls give plenty of opportunities to spot an abundance of birdlife. Adrenaline junkies can forego the birdsong for white-water rafting on the nearby Rangitata River. *www.doc.govt.nz; 2hr by car.*

10 Mt Sunday
Mt Sunday was named after boundary riders from nearby stations, who met there on Sundays. Many years later, the hill became the site of Edoras, the seat of the fictional kingdom of Rohan, during the filming of *The Lord of the Rings: The Two Towers*. Fans can tread the fields of Rohan on tours departing Methven; alternatively, the Mt Sunday track leads trampers on a one-hour circuit. *www.doc.govt.nz; 2hr by car.*

11 Kaikoura
Thanks to a particular confluence of continental shelf and current conditions, the pretty coastal town of Kaikoura is renowned for its sea-life-spotting potential. Head out on a wildlife cruise and, depending on the time of year, you might spot any number of different whale species, including humpback, pilot, sperm, southern right or killer whales, as well as dolphins, penguins and seals. *www.whalewatch.co.nz; 2hr 30min by car or 3hr by train.*

© Kevin Wells Photography / Shutterstock

——— THREE HOURS FROM ———

12 **Aoraki Mackenzie International Dark Sky Reserve**

Experience the full glory of the Milky Way in Mackenzie Country near Lake Tekapo. The entire 4300 sq km Mackenzie Basin was designated an official Dark Sky Reserve in 2012, completely free of light pollution. Numerous operators run stargazing tours, but if you're on a budget just park yourself at many of the campsites scattered across the region and wait for the sun to go down. *www.darkskyreserve.org.nz; 3hr by car.*

11

© chrisadam / Getty Images

FREEDOM CAMPING

Canterbury's dramatic scenery and abundance of pristine wilderness makes it a prime location for freedom camping – that is, camping for free on public conservation land. Keep in mind these tips when planning your responsible camping adventures:
• Look for signs noting where you can and can't camp, and what equipment you need to have (in some places, camping is only permitted for self-contained vehicles).
• Check with the local i-Site or DOC office for local campsites in your area.
• Remember the Department of Conservation's mantra: 'Carry in, carry out. Leave no trace.'
www.doc.govt.nz/ freedomcamping

12

© Lingxiao Xie / Getty Images

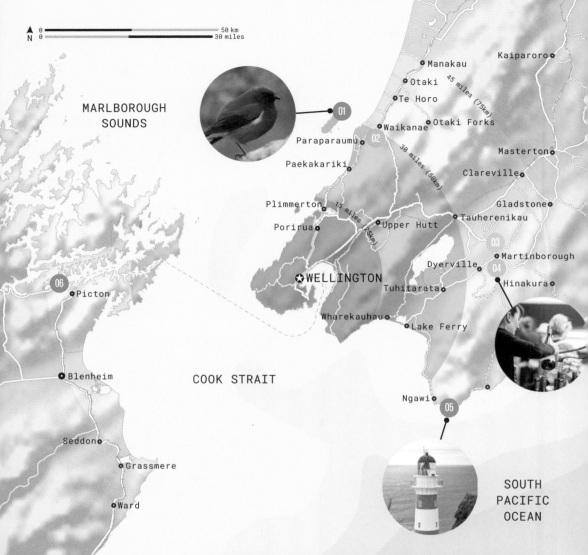

N
0 ——————————————— 50 km
0 ——————————————— 30 miles

MARLBOROUGH
SOUNDS

01

Kaiparoro

Manakau

Otaki

Te Horo

45 miles (75km)

Otaki Forks

Paraparaumu

Waikanae

02

Masterton

Paekakariki

30 miles (50km)

Clareville

Plimmerton

15 miles (25km)

Gladstone

Porirua

Upper Hutt

Tauherenikau

03

Martinborough

Dyerville

04

Hinakura

WELLINGTON

Tuhitarata

Wharekauhau

Lake Ferry

06

Picton

Blenheim

COOK STRAIT

Ngawi

05

Seddon

South
Pacific
Ocean

Grassmere

Ward

From New Zealand's compact capital, head north for culinary festivals and
vineyards, or south by ferry to one of the country's Great Walks. Nearer to town,
spot native island birds or gawp at a fascinating showcase of automotive history.

WELLINGTON

ARTS & CULTURE   HISTORY   OUTDOORS   FOOD & DRINK   FESTIVALS & EVENTS   MUSIC & FILM

——— ONE HOUR FROM ———

01 Kapiti Island Nature Tours
Punctuating the western horizon north of Wellington, Kapiti Island has been a protected wildlife reserve since 1987. Birdlife is a key reason to visit, and the island haven is home to several avian species now extinct on the mainland. Excursions with Kapiti Island Nature Tours include immersion in the country's Māori traditions, and NZ's iconic kiwi is sometimes seen on overnight stays. *www.kapitiisland.com; 1hr by car.*

02 Southward Car Museum
Regarded as one of the world's best displays of antique and unusual cars, this museum on the Kapiti Coast offers plenty of four-wheeled diversions. Among the 400 or so wonderful, historic and often surprising vehicles are a *Back to the Future* DeLorean, a Cadillac cabriolet that belonged to 1930s movie star Marlene Dietrich, and an 1895 Benz Velo, considered one of the world's first cars. *www.southward carmuseum.co.nz; 1hr by car.*

03 Cycle the Vines
Make the journey across the rugged Rimutaka Ranges and combine virtue and vice on a self-guided cycling tour around the vineyards of the Martinborough region. Local tourist information centres can provide maps, and you'll find the terrain, winding through wineries, cafes and olive groves, largely flat and easy-going. Renting a bike and stocking up on some of the planet's finest Pinot Noir is virtually mandatory. *www.wairarapanz.com/see-and-do/ cycling/cycling-vines; 1hr 10min by car.*

04 Toast Martinborough
Wellingtonians love to eat and drink, and in mid-November their focus moves from the capital's excellent cafes, craft-beer bars and bistros to this one-day festival held across the nearby Martinborough wine region. About 10 vineyards, 20 musical acts and more than 50 different dishes to sample add up to an excellent way to celebrate the beginning of summer. *www.toast martinborough.co.nz; 1hr 10min by car.*

——— TWO HOURS FROM ———

05 Cape Palliser
One of New Zealand's more remote and sparsely populated areas, Cape Palliser is a spectacular and fascinating day trip from Wellington. Coastal vistas framed by cliffs and black-sand beaches give way to quiet fishing villages, a seal colony and the view-friendly Cape Palliser Lighthouse. Shelter from southern hemisphere winds with lunch and a beer at the rustic Lake Ferry Hotel. *www.wairarapanz.com/cape-palliser; 2hr by car.*

——— THREE HOURS FROM ———

06 Queen Charlotte Track
Navigate the stunning Marlborough Sounds by ferry, and then embark on the Queen Charlotte Track, starting near the South Island port of Picton. The track begins in the tiny coastal settlement of Anakiwa, and meanders for 70km through amazing conservation reserves. Mountain biking is also an option, and it's possible to include kayaking and boat transfers along the route. *www.qctrack.co.nz; 3h 40min by ferry.*

THE LOCAL'S VIEW

'After a hard week of brewing, it's great to have a day or two in sunny Martinborough. You can't beat a relaxing afternoon at Poppies, sipping on the wines and enjoying a bountiful platter. For an excellent meat pie experience, head to the OMG cafe in the centre of town. A 40-minute drive southeast are the Putangirua Pinnacles in Aorangi Forest Park. Incredible rock formations (hoodoos) tower above a beautiful valley walk. You may recognise them from *The Lord of the Rings*, where they featured as the Paths of the Dead.'

Kelly Ryan, Fork & Brewer

© rfranca / Shutterstock; © JET PRODUCTIONS NZ; © Johan Larson / Shutterstock

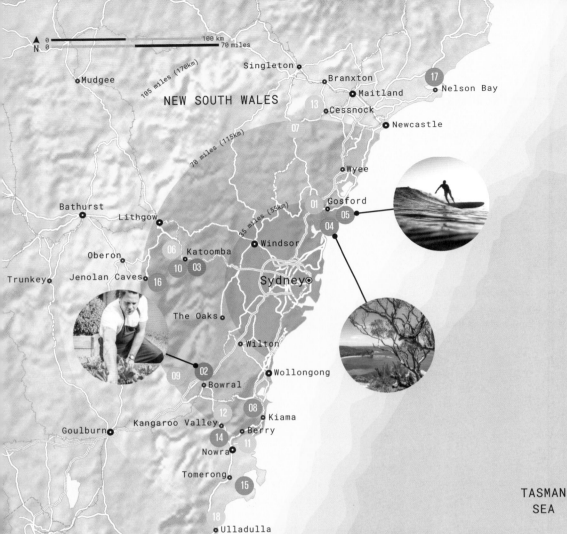

N

0
0
100 km
70 miles

Mudgee

Singleton

Branxton

17 Nelson Bay

NEW SOUTH WALES

Maitland

13 Cessnock

105 miles (170km)

07

Newcastle

70 miles (115km)

Wyee

35 miles (55km)

01 Gosford

05

04

Bathurst

Lithgow

Oberon

06 Katoomba

Windsor

10 03

Jenolan Caves

16

Sydney

Trunkey

The Oaks

Wilton

Wollongong

09 02

Bowral

The Oaks

12 08 Kiama

Kangaroo Valley

14 Berry

Goulburn

Nowra

11

Tomerong

15

18

Ulladulla

TASMAN
SEA

Drive one way from Sydney and you'll find some of the country's best surf beaches. In the other direction is the tall timber of the Australian bush. There are also wine regions, historic towns, deep caves, mighty blowholes and plenty more.

SYDNEY

● ARTS & CULTURE   ● HISTORY   ● OUTDOORS   ● FOOD & DRINK   ● FESTIVALS & EVENTS   ● MUSIC & FILM

———— ONE HOUR FROM ————

01 Gosford Classic Car Museum, West Gosford

This fairly new addition to the museum scene (it opened in 2015), is considered to be one of the best automobile museums in the world. In fact, the museum houses the largest classic car collection in the southern hemisphere, comprising more than 400 cars and motorbikes. Many of the classic cars are for sale too – just in case you're looking. *www.gosfordclassiccars.com.au; 9am–5pm Fri–Sun, by appointment Mon–Thu; 1hr 10min by car.*

02 Bowral

The scenic Southern Highlands region is home to historical little towns dotted among rolling hills and deep valleys. Bowral, the largest of these, is a place of pretty gardens, inviting wineries and restaurants such as the award-winning Biota, which lures Sydneysiders all the way from the city, just for dinner. Flora lovers should time their visit for Bowral's Tulip Time festival, which is one of the country's best flower festivals, held in September each year. *1hr 20min by car.*

03 Leura

This historic town is one of the Blue Mountains' gems. Stroll down the main street to soak up the quirky-cool vibe, stopping in at eccentric cafes for coffee and cake, and perusing boutiques for knick-knacks. One of the must-visits is Bygone Beautys Treasured Teapot Museum & Tearooms, where you can ogle the world's largest private tea collection. *www.bygonebeautys.com.au; 1hr 30min by car.*

04 Bouddi

Block off a few hours to complete this scenic 8.5km one-way walk in Bouddi National Park. Starting at Putty Beach and finishing up at Macmasters Beach, the trail tracks through rainforest terrain, along boardwalk, and past lookouts that offer sweeping views out to sea. During spring the plains are often carpeted in wildflowers; humpback whales can be seen between May and July; and in summer a swim en route at Maitland Bay comes highly recommended. *www.nationalparks. nsw.gov.au/things-to-do/walking-tracks/ bouddi-coastal-walk; 1hr 30min by car.*

05 Central Coast

Australia is home to hundreds of surf beaches, and some of the best ones are located on the Central Coast. For beginners, the small, long waves at Umina Beach and Terrigal Beach make for ideal learning conditions. More experienced surfers will find decent breaks at Macmasters Beach, Killcare, Avoca Beach and Copacabana. *www. visitcentralcoast.com.au; 1hr 30min by car.*

———— TWO HOURS FROM ————

06 The Wintergarden at Hydro Majestic Hotel

Feast on contemporary Australian cuisine or enjoy decadent high tea while gazing at views of Megalong Valley through panoramic windows. This elegant restaurant is housed in one of the grand original hotels in the Blue Mountains. Have a nose around after you dine to admire the design details. *www.hydromajestic.com.au/dining/ the-winter-garden; 1hr 40min by car.*

BLUE MOUNTAINS HIKING

The Blue Mountains National Park is home to hundreds of walking trails, ranging from easy short walks to wheelchair-accessible tracks to challenging multi-day hikes. It's important to research and prepare. Ensure you always carry plenty of water and some snacks; wear appropriate hiking shoes; walk in groups; assess the difficulty level of the trek before setting out; and advise a family member or friend about your walking plans. If you're camping, more extensive preparation is needed. Check www.nationalparks.nsw.gov.au for updates and advice.

© Pete Seaward

07 Sculpture in the Vineyards

The historic village of Wollombi in the Hunter Valley (along with some of the surrounding vineyards) is transformed into a series of fascinating arts sites every November. Unique modern sculptures created by Australian artists emerge, and tours, talks and walks are all part of the imaginative programme. Wine tasting and buying also play central roles in the event – this is the Hunter Valley, after all. *www.sculptureinthevineyards.com.au; Nov; 1hr 40min by car.*

08 Kiama Blowhole

One of the world's largest blowholes, located in Kiama, makes quite the impression on passers-by. Time it right (it comes down to sea conditions and patience), and a profusion of seawater will erupt into the air, followed by a thundering sound. Watch the spectacle over and over again, then head to Kiama Beach for a swim, take in the views of Kiama Lighthouse, and visit Pilot's Cottage Museum. *www.kiama.com.au; 1hr 40min by car.*

09 Joadja

A few empty homes and dilapidated buildings make up this once-flourishing village, but the lack of habitable housing and people is the attraction. Formerly comprising a population of approximately 1000 (predominately Scottish migrants who worked as shale miners), Joadja is now a ghost town, only opened up to visitors a few times a year (or by private appointment) for heritage tours. There's a whisky distillery there too, which also has restricted opening hours. *www.joadjatown.com.au; 1hr 40min by car.*

10 Three Sisters

These jagged sandstone formations known as the Three Sisters are one of the Blue Mountains' most famous sights, towering over Jamison Valley and making for one unusual and spectacular scene. The best view is from Echo Point lookout, near the bustling town of Katoomba. There are also a number of hiking trails where you can admire views of the sisters, including a short 0.8km round trip, the Three Sisters walk. *www.nationalparks.nsw.gov.au/things-to-do/lookouts/echo-point-lookout-three-sisters; 1hr 50min by car.*

11 Berry

Grass-carpeted hills and sprawling valleys surround the charming town of Berry. Stop at the famous Berry Sourdough Bakery for coffee and croissants, then browse the boutiques and local produce stores along the main street. If hunger calls again, there are a number of fantastic restaurants serving regional produce. South on Albany and Hungry Duck are top-notch. *www.berrysourdoughcafe.com.au; www.southonalbany.com.au; www. hungryduck.com.au; 1hr 50min by car.*

 Kangaroo Valley Folk Festival
Every October, various venues around Kangaroo Valley are transformed into festival spaces and, for three fun-filled days, the valley comes alive with the sound of folk music. There are concerts and dance workshops, t'ai chi sessions and poetry recitals, arts, crafts and special events for kids. It's a ticketed event, but with more than 100 concerts and talented performers, from Australia and overseas, it's money well spent. *www.kangaroovalleyfolkfestival.com.au; Oct; 2hr by car.*

13 Pokolbin
This friendly village in the Hunter Valley, Australia's oldest wine region, is a perfect base for wine aficionados. Some of the region's top wine producers, such as Audrey Wilkinson and Cockfighter's Ghost, are located in Pokolbin, and there are plenty of others nearby. Beyond wine tasting, the town also has a number of accommodation options, produce shops and restaurants. *www.audreywilkinson.com.au; www. cockfightersghost.com.au; 2hr by car.*

THE ENTRANCE PELICAN FEED

It began accidentally over 20 years ago, when staff at Clifford's fish and chip shop in The Entrance town (1hr 30min by car), would toss leftover scraps to the local pelicans, which would gobble them up enthusiastically. Jimbo's Quality Seafood took over and from there the event grew until it became legendary around the state. Today it's a daily affair, sponsored by a number of local businesses, and regularly brings in the crowds. At 3.30pm every day, a flock of pelicans will jump and screech for their daily fishy treat – bring your camera. *www. theentrance. org.au/explore/ pelican-feeding; 3.30pm daily.*

© Kok Kai Ng / Getty Images

© Jonathan Stokes

© totajla / Shutterstock

14 Kangaroo River

Drive for no more than a couple of hours and you're well out of the Big Smoke and in the middle of nature's wonderland. Kangaroo Valley is a great destination for outdoor activities, including canoeing and kayaking the gentle rapids of Kangaroo River. Several tour companies rent out canoes and kayaks from various start and end points. The historic suspension bridge is one of the most popular sights of note. *2hr 10min by car.*

15 Jervis Bay

White-sand beaches that stretch as far as the eye can see and sun-shimmering turquoise water are the norm in Jervis Bay, with beautiful Hyams Beach a particular favourite. Swimming, kayaking and watching dolphins frolic in the sea is time well spent. Or find a patch of grass and simply kick back with a good book – and an even better view. *2hr 30min by car*.

—— THREE HOURS FROM ——

16 Jenolan Caves

An incredible underground adventure awaits those keen to explore Australia's most famous caves – a stunning labyrinth of more than 300 stalactite-lined limestone grottos. Nine of the caves are open to visitors, and there are a variety of tours seven days a week. Guided walking options are the most popular, but adventure caving and night tours on Fridays and Saturdays are exciting alternatives. *www.jenolancaves.org.au; 2hr 40min by car.*

Nelson Bay

This idyllic coastal town in Port Stephens comes alive in the warmer months when holidaymakers flock to the sea. Swimming is a popular pastime; there's excellent snorkelling and diving on offer; and the area is known for its ideal sailing conditions – and its abundance of delicious seafood restaurants. From May to November, Nelson Bay is a prime location for whale watching. *2hr 40min by car.*

Rick Stein at Bannisters

Now one of Australia's pre-eminent celebrity chef-affiliated restaurants, Rick Stein at Bannisters has been in business since 2009. Located in Mollymook's Bannisters by the Sea luxury lodge, the restaurant showcases Stein's signature ingredient – seafood. The fantastic ocean views are best enjoyed while it's light, so go for lunch or early dinner. *www.bannisters. com.au/rick-stein; 3hr by car.*

CULTURAL EXPERIENCE WITH NURA GUYU

Learn about Indigenous Australians and their history on the South Coast – and discover dozens of bush food plants – by joining a half- or full-day cultural learning experience with Nura Guyu (*nuragunyu.com. au; 3hr 10min by car to Ulladulla*). Budawang elder Noel Butler, from Yuin nation, shares his extensive knowledge of traditional culture at these events, and works alongside his wife Trish Butler to teach attendees about the value of the land. Bush-tucker walks (held inland or on the coast), food preparation and tasting, dance, and discussions about arts and artefacts are all part of the day.

© Na Gi Sa Ikeda / Getty Images

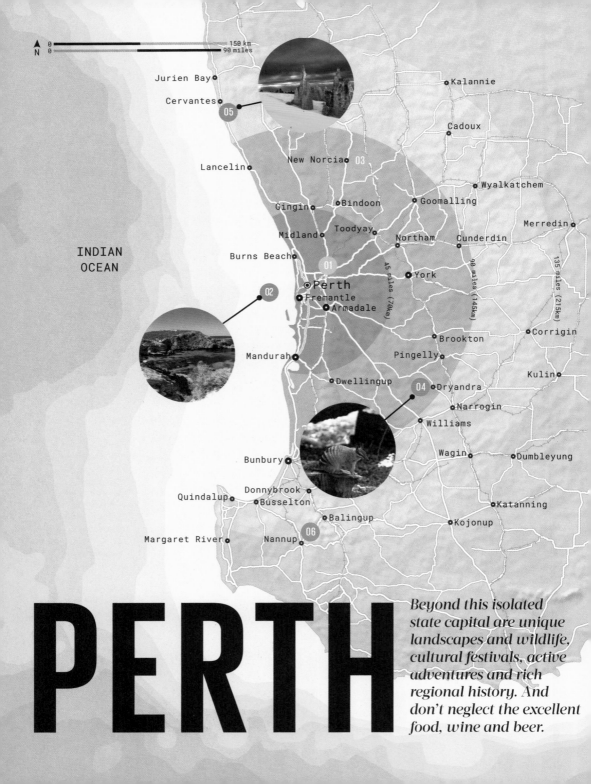

N
0 ——— 150 km
0 ——— 90 miles

INDIAN
OCEAN

Jurien Bay
Cervantes
05
Lancelin
New Norcia **03**
Gingin
Bindoon
Goomalling
Midland
Toodyay
Northam
Cunderdin
Burns Beach
01
Perth
Fremantle
York
Armadale
Mandurah
Dwellingup
04 Dryandra
Narrogin
Williams
Bunbury
Donnybrook
Quindalup
Busselton
Balingup
Kojonup
Margaret River
Nannup **06**
Brookton
Pingelly
Kulin
Wagin
Dumbleyung
Katanning

Kalannie
Cadoux
Wyalkatchem
Merredin
Corrigin

45 miles (70km)
90 miles (145km)
135 miles (215km)

PERTH

Beyond this isolated state capital are unique landscapes and wildlife, cultural festivals, active adventures and rich regional history. And don't neglect the excellent food, wine and beer.

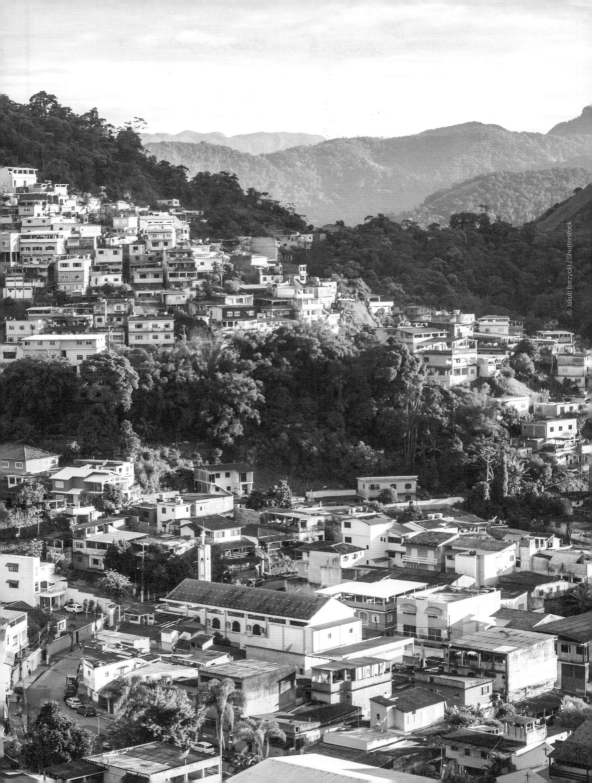

N 0 _____ 100 km
0 _____ 70 miles

BRAZIL

MINAS GERAIS

Leopoldina

Minduri

Juiz De Fora

RIO DE JANEIRO

105 miles (170km)

Caxambu
Liberdade

Bom Jardim De Mines

Rio Preto

Conceicao De Macabu

70 miles (115km)

Valenca

Nova Friburgo

Macaé

Resende

Barra Do Pirai

05

Teresopolis

08

SÃO PAULO

Barra Mansa

Petropolis

07

Guapimirim

06

Silva Jardim

14

35 miles (55km)

Rio Bonito

10

15

Rio de Janeiro

02

Niterói

Araruama

Cabo Frio

01

11

12

13

09

04

03

16

17

Paraty

18

ATLANTIC OCEAN

Sure, there's history and culture to find around Rio. But you're here for sun, sand and cachaca. Luckily, you can have it all: close to the city are dazzling beaches, historic hikes and the site where Amerigo Vespucci 'discovered' South America.

RIO DE JANEIRO

● ARTS & CULTURE ● HISTORY ● OUTDOORS ● FOOD & DRINK ● FESTIVALS & EVENTS ● MUSIC & FILM

——— ONE HOUR FROM ———

01 Museu de Arte Contemporânea de Niterói

The spaceship-like, Oscar Niemeyer-designed building that houses the Museu de Arte Contemporânea de Niterói (MAC) is a destination in itself. In addition to an impressive collection of modern Brazilian art, the museum has a special gallery with spectacular views of Sugarloaf Mountain and Guanabara Bay. To get there, you'll cross the Rio–Niterói bridge – one of the longest in the world. *www.culturaniteroi.com.br/macniteroi; 40min by car or 1hr 30min from Copacabana on bus 740.*

02 Mercado São Pedro, Niterói

After a visit to the Niterói Contemporary Arts Museum (MAC), have lunch at Niterói's colourful marketplace. Traditional and practically tourist-free, the Mercado São Pedro specialises in fresh seafood. Take a stroll around the lower level to check out the catch of the day, then head upstairs for a humble feast in one of several no-frills, family-run restaurants. *www.mercadodepeixesaopedro.com.br; 40min by car or 1hr 30min from Copacabana on bus 740.*

03 Sitio Burle Marx

The former home of the Brazilian landscape architect Roberto Burle Marx is his magnum opus – and it's been recognised by Unesco for its cultural value. The main attractions are the sprawling gardens and multiple reflecting pools: here, you'll see one of the world's largest collections of tropical and subtropical plants. A guided tour takes you inside, too, to see the architect's innovative drawings and plans. *www.visit.rio/en/que_fazer/sitio-burle-marx-2; 1hr by car or 2hr by bus.*

04 Ilha de Itacuruçá

One of a cluster of three islands close to Rio, Itacuruçá hardly sees any international tourists, which is, of course, part of what makes the place so peaceful. Sunbathe on Praia Grande beach – or catch a water taxi to a quieter stretch of coastline – hike to the Itinguçú waterfalls, or dine on freshly caught seafood at a laid-back beach bar. *1hr 30min by bus or car, plus a short boat ride.*

05 Petrópolis

Sitting on the edge of Serra dos Órgãos National Park, Petrópolis has long been a favourite weekend escape for cariocas (Rio residents). But the 'Imperial City', north of Rio, is also a destination for history buffs: points of interest include the stately São Pedro de Alcântara cathedral and a museum housed in the one-time palace of a 19th-century emperor. *1hr 30min by car or 1hr 30min by bus from Novo Rio Rodoviaria.*

——— TWO HOURS FROM ———

06 Casa Stefan Zweig

In the 1920s, the Viennese novelist, playwright and journalist Stefan Zweig was one of the most famous writers in the world. During Hitler's rise to power in Europe, he fled Austria with his wife, establishing residence in Petrópolis, Brazil in 1942. Sadly, the couple died by joint suicide just five months later. Casa Stefan Zweig is now a

DANCING IN THE STREET

Experiencing Carnaval (sometimes spelled 'Carnival') at the Sambadrome in Rio is a once-in-a-lifetime thrill. But you need to plan well ahead to get your hands on tickets and book accommodation. Don't despair if you miss out – you'll find smaller but equally vibrant February celebrations in many of the cities and communities around Rio, particularly along the coast, where street parades and revelry are more accessible. Arraial do Cabo and Cabo Frío, each now hosting high-spirited festivals, are particularly popular destinations during Carnaval time. If you'd rather skip all the street parades, head to quieter Búzios.

museum dedicated to the author's literary legacy. ***www.casastefanzweig.org;*** *1hr 40min by car.*

07 **Serra dos Órgãos National Park**
Featuring massive rocks that jut dramatically into the sky – the park's name, chosen by early Portuguese explorers, refers to these unusual formations, comparing them to organ pipes – Serra dos Órgãos National Park is a dream destination for avid rock-climbers and hikers. With waterfalls, well-marked trails and infrastructure for campers, it's a beautiful place to spend a weekend in the wilderness.
www.parnaso.tur.br; *1hr 40min by car.*

© Brazil Photos / Getty Images

——— THREE HOURS FROM ———

08 **Aldeia Velha**
Get away from it all in Aldeia Velha, a small village surrounded by thick forests and lush waterfalls in a remote corner of Rio de Janeiro state. Go for a self-guided hike, or take off on a horseback-riding adventure, then join the locals – the village has just 800 residents – for a cold beer in one of Aldeia Velha's casual outdoor cafes and bars. *2hr 40min by car.*

09 **Angra dos Reis**
Many travellers blow right past Angra dos Reis on their way to Ilha Grande – the coastal town is a gateway to the island. But it's worth slowing down and spending time in both. Angra dos Reis means 'Bay of Kings' in Portuguese, and the town's location is regal indeed: on one side, it's bordered by a bright blue bay dotted with 365 islands, and on the other by a hilly forested

landscape that is laced with hiking trails. *2h 40min by car or 3h by bus.*

10 **Cabo Frio**
Cabo Frio is one of the oldest cities in Brazil. Explore more than 500 years of history at Forte São Mateus, a one-time stronghold against pirates. Then relax at one of Cabo Frio's many lovely white-sand beaches – popular Praia do Forte, with clear blue water, is more than four miles long. Or take a schooner ride to discover the rocky coastline from a different perspective. *3hr by car or bus.*

11 **Arraial do Cabo**
Arraial do Cabo, protruding into the Atlantic Ocean, was the landing place of Amerigo Vespucci in 1503. Countless shipwrecks by Portuguese, French, Dutch, English and Brazilian fleets have occurred just off the coast, some of which are documented in a small marine museum. Today, the town,

complete with its lively seafood market, still has the feel of a fishing village. *3hr 10min by car or bus.*

12 **Arraial do Cabo beaches**
It's no surprise that Arraial do Cabo is a favourite destination for Cariocas. Surrounded by sand dunes and facing the pretty landscapes of Ilha do Farol, the place has breathtaking white beaches, such as postcard-perfect Pontal do Atalaia. The green water is cold, making this a popular spot for scuba divers and whale-watching enthusiasts – migrating humpback whales pass directly offshore. *3hr 10min by car or bus.*

13 **Ilha do Farol**
A tiny island just off the coast, a stone's throw from Arraial do Cabo, the so-called 'Lighthouse Island' is worth the trip for its historic lighthouse and a biological reserve that's home to diverse wildlife. Two

other lighthouses stood here before the current incumbent, built in 1926 and managed by the Brazilian Navy. *3hr 10min by car or bus, plus a short boat ride.*

14 **Búzios**
There are lots of reasons to make the trip to Búzios, the Brazilian resort town popularised by Brigitte Bardot in her heyday. As well as picturesque beaches – 17 of them, to be exact – surf breaks, and a horseshoe bay that's ideal for stand-up paddleboarding and other water sports, Búzios has a lively dining and drinking scene centred around cobblestoned Rua das Pedras. *3hr 20min by car or bus.*

15 **Praia de Ferradurinha, Búzios**
There are lots of great beaches in Búzios, but lovely Praia de Ferradurinha, tucked between busier stretches of coastline, is something of a hidden gem. Locals debate whether it's a proper beach or

GATEWAY TO GETAWAY

If you're taking the bus to a destination close to the city of Rio – Búzios and Niterói are popular options – you'll need to go to Novo Rio bus station in the downtown area of the city. Dozens of bus companies operate out of the terminal, and the station can be chaotic at peak travel times. It's also notorious for petty theft, so pay close attention to your valuables. Visit the terminal website at www. transportal.com. br/rodoviaria-novorio for more information.

09

© Jakub Barzycki / Shutterstock

a natural swimming pool between two cliffs – either way, it features some of the clearest water along these shores, a curving sandy beach and colourful rock formations.
3hr 20min by car or bus.

16 Ilha Grande

Getting to the beautiful island of Ilha Grande, which is famous for its idyllic white sandy beaches (and the absence of both roads and cars), is an adventure. You'll take a bus, then a ferry that leaves from the ports of Angra dos Reis or Mangaratiba, both located west of Rio – or you could book a transfer service that covers the entire journey. It's well worth the effort.
3hr 30min: 2hr 30min by bus plus 1hr-1hr 45min by boat.

© Stefano Paterna / Alamy Stock Photo

© OSTILL is Franck Camhi / Shutterstock

© Deni Williams / Shutterstock

Lazareto, Ilha Grande

These eerie ruins on Ilha Grande are all that's left of a quarantine-style hospital built in 1871 to house sick passengers arriving during the first waves of immigration into the country. Later, Lazareto functioned as a prison. See the ruins on the short Circuito do Abraão hike, which takes you past a 19th-century aqueduct and a swimming hole where you can take a refreshing dip. ***3hr 30min: 2hr 30min by bus plus 1hr–1hr 45min by boat.***

Paraty distilleries

As if the picturesque Unesco-honoured city of Paraty wasn't appealing enough on its own, it happens to be set in the middle of one of Brazil's traditional cachaca-producing regions. The potent rum-like spirit is made from sugar cane, and it's the basis for the caipirinha, the country's beloved national cocktail. Take a distillery tour or indulge at one of Paraty's lively bars, where the tables spill out onto the cobblestoned streets. ***4hr by car.***

TOP WILDLIFE-VIEWING DESTINATIONS IN LATIN AMERICA

Monkeys and sloths and condors... Up in the mountains, on a boat or underwater, inspiring wildlife experiences abound outside Latin American cities.

Monkeys, Las Isletas, Nicaragua

You'll spot them from the water as your wooden boat cruises around Las Isletas, a group of small islands in Lake Nicaragua. One of the 365 islets is inhabited only by monkeys that swing through the jungle trees.
30min from Granada or 1hr 20min from Managua by boat tour.

Penguins, Isla Martillo, Argentina

One of the largest penguin colonies on the planet is at the end of the world – in Tierra del Fuego, near Antarctica. Between November and March, marvel at Magellanic penguins swimming and playing on Martillo Island.
4hr from Ushuaia by boat.

Sloths, Costa Rica

You're guaranteed to get a close look at these slow-moving tree-dwellers at the Sloth Sanctuary on the Caribbean coast, dedicated to sloth rescue and rehabilitation.
www.slothsanctuary. com; Limón province; Tue-Sun; US$30; 30min from Limón or 4hr from San José by car.

Tropical Fish, Búzios

The clear turquoise waters off the coast of Búzios are a natural aquarium for colourful tropical fish and perfect for snorkelling. Access the water near the cliffs of João Fernandes, João Fernandinho, Forno or Azeda.
3hr 20min from Rio de Janeiro by car or bus.

Sea Turtles, Brazil

Projeto TAMAR is a Brazilian non-profit organisation that works to protect sea turtles from extinction. There are several locations to visit along the coastline: a great option is the recovery centre at Praia do Forte.
www.tamar.org; US$6; 1hr 20min from Salvador de Bahia by car or bus.

Whale-watching, Argentina

Put to bed any memories of disappointing whale-watching excursions. In the waters off the coast of Puerto Pirámides, whales swim around your boat, breach, and slap the water with their fins and tails.
www.puertopiramides. gov.ar; Jun-Dec; 1hr from Puerto Madryn by car or bus.

Hummingbirds, Colombia

Just an hour or two outside Bogotá (in any direction), get ready to spot exotic birds in the cloud forest or dozens of varieties of hummingbirds. The guides at Birding Bogotá & Colombia can plan your dream birdwatching excursion.
www.birdingbogotaand-colombia.com; costs vary; 1hr from Bogotá by car.

Llamas, Machu Picchu, Peru

What could make the view of Machu Picchu even more memorable? Seeing the lost city of the Incas with a couple of photogenic llamas and alpacas nearby, that's what. Both live in the wild here in the Andes.
6am-4pm daily; US$47; 3hr from Cusco by train.

Condors & Pumas, Torres del Paine National Park, Chile

In the stark, wide open spaces of Torres del Paine in southern Chile, you'll see condors flying overhead – and perhaps a puma on the trail.
www.torresdelpaine. com/en; open daily; US$22; 1hr 30min from Puerto Natales by car or bus.

Whale Sharks, Utila, Honduras

Swimming alongside a whale shark – the biggest fish on the planet – is no fantasy in the clear waters around the Bay Islands. Mid-February to April is your best bet, but you could see the massive fish any time of the year.
1hr from La Ceiba by ferry.

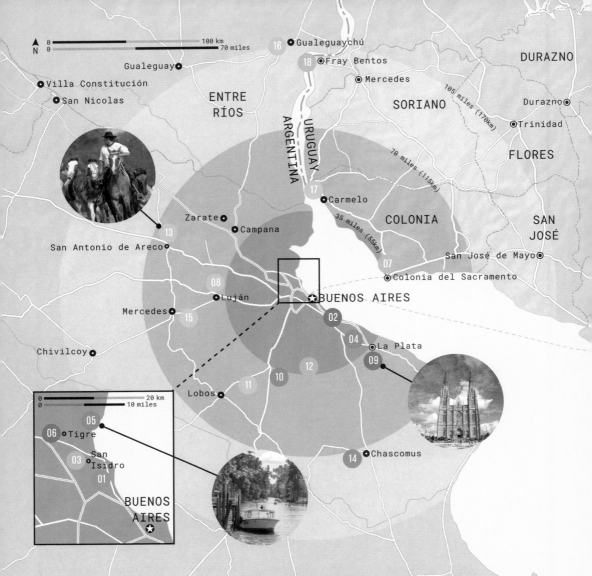

N

0 ————————— 100 km
0 ————————— 70 miles

Gualeguaychú 16

Fray Bentos 18

DURAZNO

Gualeguay

Mercedes

Villa Constitución

SORIANO

Durazno

San Nicolas

ENTRE RÍOS

105 miles (170km)

Trinidad

FLORES

Carmelo 17

COLONIA

70 miles (115km)

SAN JOSÉ

Zarate

Campana

35 miles (55km)

San José de Mayo

13

San Antonio de Areco

08

07

Colonia del Sacramento

BUENOS AIRES

Luján

02

Mercedes 15

04

La Plata

Chivilcoy

12

09

0 ————— 20 km
0 ————— 10 miles

10

05

11

Lobos

06 Tigre

03

San Isidro

01

14 Chascomus

BUENOS AIRES

Argentinians often say Buenos Aires is a bubble, separated from the rest of the country. Burst that bubble and you can explore the rural pampas, the festivals and thriving arts scenes of nearby cities, and then hop on a ferry to Uruguay.

BUENOS AIRES

● ARTS & CULTURE ● HISTORY ● OUTDOORS ● FOOD & DRINK ● FESTIVALS & EVENTS ● MUSIC & FILM

——— ONE HOUR FROM ———

01 Peña Circo Criollo, Olivos

Buenos Aires is known as the birthplace of the melancholic tango, but breach the city limits and you'll discover that the rest of the country dances to a more upbeat tune. Folklore (folk music and dance) is celebrated at *peñas*, such as the Peña Circo Criollo in Olivos. Here, live folk bands play and revellers dance *chacareras*, featuring rhythmic handclapping; and *zambas*, in which dancers pass each other while waving handkerchiefs. *www.facebook.com/ pcircocriollo; 30min by car or 1hr by bus.*

02 Quilmes

Fans of street art should head to the town of Quilmes in Buenos Aires' southern suburbs. As part of a project called Pinta tu Barrio (paint your neighbourhood), approximately 10,000 sq metres of exterior walls have been brightened by murals drawn by local and international artists, including renowned Argentinian muralists Martín Ron and Milu Correch. *30min by train from Constitución.*

03 San Isidro

The salubrious suburb of San Isidro is steeped in history. Revolutionary hero General Juan Martín de Pueyrredón and José de San Martín planned battle tactics against the Spanish at Pueyrredon's home, now the Museo Pueyrredon, while Quinta los Ombúes is the historic villa where the Argentinian national anthem was supposedly first sung. Nearby Villa Ocampo is the impressive former mansion of writer Victoria Ocampo. *50min by train from Retiro.*

04 Caballos a la Par

Channel your inner gaucho at Caballos a la Par. These exceptional stables have developed a winning technique to teach wannabe riders: the key is riding *a la par* (side by side) with the instructor. Guided rides are given in the woods of a provincial park near La Plata; transport from Buenos Aires is provided. Even if you've never ridden before, you might be cantering by sundown. *www.caballos-alapar.com; 50min by car.*

05 Tigre Delta

The town of Tigre is the gateway to the Paraná Delta, where muddy waterways wind around lush, green islands, beneath overhanging branches and past rickety wooden jetties that extend from stilted houses. The best way to explore it is by kayak, paddling around until the whiff of barbecue invites you to pull up at a sandy shore for lunch at a low-key restaurant. *http://eldoradokayak.com; 50min by car or 1hr by train from Retiro.*

06 Museo de Arte Tigre

Perched on a riverbank in glorious grounds, Tigre's art museum showcases the work of Argentina's most celebrated artists. The stunning belle époque building was an elite social club at a time when Tigre was the weekend retreat for the rich. It's easy to imagine the parties that took place here as you ascend the grand marble staircase. *www.mat.gov.ar; 50min by car or 1hr by train from Retiro.*

07 Colonia del Sacramento, Uruguay

A favourite getaway for Porteños (people from Buenos Aires) is to jump on

VISITING AN ESTANCIA

The countryside surrounding Buenos Aires is dotted with *estancias* (ranches), many of which double as (often luxury) hotels. Most offer a day package option, which usually involves horse riding, a barbecue lunch and afternoon tea, and use of the grounds and pool. Near San Antonio de Areco, El Ombú de Areco (*www. estanciaelombu. com; 1hr 30min by car*) is one of the best; activities include demonstrations of traditional horse riding by gauchos. Other good *estancias* nearby are Los Dos Hermanos in Zarate (*www. estancialosdos hermanos.com; 1hr 20min by car*) and La Candalaria in Lobos (*www. estanciacan delaria.com; 1hr 30min by car*).

© evenfh / Shutterstock

a boat and cross the Río de la Plata to Colonia del Sacramento in Uruguay. Due to its picturesque cobbled streets and a laid-back vibe, Colonia feels a world away from the bright lights of Buenos Aires glittering across the water. The historic old town has eight small museums that can be visited on a single ticket. **1hr by boat.**

08 Luján Pilgrimage

On the first Saturday in October, throngs of the faithful walk the 60km from the Buenos Aires neighbourhood of Liniers to Luján's basilica, to pay tribute to the Virgen de Luján, Argentina's patron saint. Arriving on the Sunday of the pilgrimage weekend, you'll spot families of exhausted pilgrims snoozing in the square, enjoying riverside barbecues and filling plastic bottles with holy water from the fountain. ***Oct; 1hr by car or 2hr by bus from Plaza Italia.***

09 La Plata

Looming over the city of La Plata are the twin towers of an impressive neo-Gothic cathedral. The neoclassical facade and stained-glass windows of the Museo de La Plata are also worth a look, but the star attraction for architecture buffs is Casa Curutchet, a house designed by Le Corbusier and his only completed work in Latin America. ***www.capbacs.com/ capba-casa-curutchet; 1hr by car or 1hr 10min by train from Constitución.***

10 Puesto Viejo

Leave the city behind and enter a world of polo ponies, elegant country grounds, infinity pools and indulgent

afternoon teas, if only briefly. A day at Puesto Viejo Estancia involves polo lessons, a barbecue lunch and the chance to watch professionals play a four-chukka match. If, come nightfall, you can't tear yourself away, why not bed down at the luxurious hotel. ***www.puestoviejoestancia.com.ar; 1hr 10min by car.***

11 Fiesta de la Picada y la Cerveza Artesanal, Uribelarrea

In November, craft-beer brewers gather in the village of Uribelarrea in celebration of *cerveza artesanal*, and more than 30 local producers set up stalls. In Argentina, beer is usually accompanied by a *picada* (a sharing board of cheeses and cold cuts); this festival honours them both, with fine regional produce on offer to soak up the booze. The merriment is further enhanced by live music. ***www.facebook.com/ FiestaDeLaPicadaYLaCervezaArtesanal; Nov; 1hr 20min by car.***

12 Museo Histórico 17 de Octubre, Quinta San Vicente

Few people know about it, but just outside Buenos Aires, in San Vicente, is the former *quinta* (country house) of perhaps Argentina's most important political figures, President Juan Perón and his wife Eva. Now a museum, the house contains the couple's personal possessions, laid out as though they were about to return. Also here is the presidential train on which the Peróns travelled, and the mausoleum where Juan Perón is buried. ***www.gba.gob.ar/cultura/museos; 1hr 20min by car.***

11

© Y.Levy / Alamy Stock Photo

08

© Anton Velikzhanin / Alamy Stock Photo

FRANCISCO SALAMONE

During the 1930s, prolific architect Francisco Salamone left his mark on scores of pampas towns with his striking designs for public buildings, including town halls, plazas, cemeteries and slaughterhouses. His futurist art deco style is realised in large-scale works in rural towns across Buenos Aires province. One of the best ways to see them is on a road trip. About four hours' drive from the capital are the cemetery and plaza of Azul and Rauch town hall. An hour more will take you to the cemetery and town hall at Laprida, the town hall at Carhué and the eerie abandoned slaughterhouse at Villa Epicuén.

13 Fiesta de la Tradición, San Antonio de Areco

The fertile grasslands surrounding Buenos Aires are the home of the gaucho, a traditionally nomadic, cowboy-like figure on horseback. Modern gauchos gather in the prosperous pampas town of San Antonio de Areco every November, sporting their finest horse gear and showing off their riding skills at the Fiesta de la Tradición. The party gets into full swing with folk dancing and plenty of grilled meat. ***http://sanantoniodeareco. tur.ar; Nov; 1hr 30min by car or 2hr by bus from Retiro.***

──────── TWO HOURS FROM ────────

14 Laguna de Chascomús
The peaceful lakeside town of Chascomús is the perfect antidote to frenetic Buenos Aires. A paved path loops around the 30km circumference of the picturesque lake, making for a beautiful bike ride. Other possible pursuits include kitesurfing and windsurfing, and the stretch of shoreline close to Chascomús town is a popular place for a stroll. The sunsets over the water are spectacular. ***1hr 40min by car or 2hr 10min by bus from Retiro.***

15 Tomás Jofre
At weekends, the tiny village of Tomás Jofre near Mercedes swells with visitors who come to feast on local salami, homemade pasta and – this being Argentina – mountains of barbecued meat. This sleepy rural idyll has become known as a gastronomic hotspot, and on Saturdays and Sundays dozens of restaurants and traditional food stores buzz with diners. Come here on a sunny day and

eat your lunch in the open air, beneath the trees. ***1hr 50min by car.***

──────── THREE HOURS FROM ────────

16 Carnaval de Gualeguaychú
Feathers, glitz, scantily clad dancers, colourful costumes, elaborate floats and resonating drumbeats transform the otherwise quiet town of Gualeguaychú into Argentina's biggest carnival, held every Saturday in January and February. More than 1000 performers take part in the parade and compete to be named the winning *comparsa* (carnival troupe), while spectators work up a sweat dancing in the stands. ***www.carnavaldelpais.com.ar; Sat, Jan & Feb; 2hr 40min by car or 3hr 20min by bus from Retiro.***

© GM Photo Images / Alamy Stock Photo

17 Narbona Wine Lodge, Carmelo, Uruguay

This boutique winery and dairy farm near Carmelo is heaven for wine-lovers and foodies alike. You can work up a thirst cycling through the vineyards before enjoying a wine and cheese tasting, or have a go at creating your own blend under the guidance of an oenologist. The restaurant in the restored homestead serves gourmet pastas and local beef. We'd recommend you sleep it all off at the adjoining hotel. *www.narbona.com.uy; 3hr by boat from Tigre.*

18 Museo de la Revolución Industrial, Fray Bentos, Uruguay

A former British-owned meat processing plant might not be everyone's idea of a getaway, but the abandoned factory in the Uruguayan town of Fray Bentos makes for an offbeat but intriguing trip. The erstwhile meatpacking plant is now a museum, with colourful displays that bring the factory's history to life. Tours grant access to the maze of passageways in the abandoned slaughterhouse out the back. *www.paisajefraybentos.com/pc; 3hr 30min by car.*

ARTISAN MARKETS

At weekends, several leafy plazas on the outskirts of Buenos Aires and in nearby towns host craft markets, where artisans set up stalls and sell their wares. Handcrafted items to look out for include *mates* (gourds for drinking a local type of tea) and jewellery. Not to be missed is the Sunday folk fair in the neighbourhood of Mataderos, on the western edge of the city (1hr from downtown BA by bus), which also features live folk music, regional cuisine and even horse riding displays. Other weekend artisan fairs are held in San Telmo (*left*), La Plata and San Pedro.

© tateyama / Shutterstock

N
0 — 200 km
0 — 110 miles

Huarmey◉

ANCASH

◉Huánuco

HUÁNUCO

Puerto
Bermúdez◉

165 miles (265km)

PASCO

Barranca◉

◉Cerro de Pasco

PERU

110 miles (175km)

05

Huaura◉

Junín◉

◉San Ramón
Satipo

LIMA

◉Tarma

55 miles (90km)

◉La Oroya

Chancay◉

JUNÍN

Jauja◉

◉Chosica

Callao◉ ✪LIMA

◉Huancayo

Chilca◉ 01

Huancavelica◉

03

HUANCAVELICA

Chincha Alta◉ 02

PACIFIC
OCEAN

Pisco◉

Paracas◉ 04

ICA

Huacachina◉ 06 ◉Ica

◉Santiago

AYACUCHO

LIMA

After exploring the cosmopolitan
city, a short hop will get you into
the desert, out to sea and on to
new adventures. Peru's coastline
near Lima is a ghostly expanse of
brown beaches, working fishing
villages and desert oases.

ARTS & CULTURE　　HISTORY　　OUTDOORS　　FOOD & DRINK　　FESTIVALS & EVENTS　　MUSIC & FILM

——— ONE HOUR FROM ———

01 Pucusana
Heading south from Lima, the eternal fog of the city somehow lifts when you hit Pucusana. Much loved as a beach destination, this bustling port village also houses the largest fishing fleet in the region, and the ceviche here is out of this world. Make straight for the port to sample the freshest ceviche and seafood dishes, washed down with an ice-cold Cusqueña beer – you'll think you've found heaven. *1hr by car or bus.*

02 Chincha
A stronghold of Afro-Peruvian music and culture, Chincha moves to a different beat than most of Peru. The nearby village of El Carmen is famous for its live music and dance exhibitions. Expect syncopated beats from wooden crates (known as the cajon), buzzing from the *quijada* (donkey jawbone), and flamenco-styled Zapateo dances. Nearby, explore a few worthwhile archaeological sites at Tambo de Mora and La Centinela. *2hr 50min by car.*

03 Lunahuaná
The Rio Cañete is the lifeblood of this wine-producing valley south of Lima, a slice of green cutting perfectly through the desert browns. The river itself provides great rafting for those adventurous souls keen to take on the thrilling Class II to IV waters (there are also a few easier runs more suitable for families). After your river run, hike up five minutes from town to the scenic mirador for views down to the valley below. *3hr by car.*

——— TWO HOURS FROM ———

04 Paracas
There are plenty of adventure opportunities on the Paracas peninsula. Make time for a half-day boat tour to the Islas Ballestas, where you can see the famous Candelabra Geoglyph, glide past cacophonous sea lion colonies and spot Humboldt penguins. On a good day, you'll see thousands of birds and maybe even a dolphin or two. Another recommended trip from here explores lost fishing villages and desolate coastal areas. *3hr by car.*

05 Caral
Set in the lush Supe Valley, 25km from Barranca, the archaeological site at Caral gives you interesting views of amphitheatres, ceremonial platforms and pyramids. It makes for a wonderful break from the chaotic energy of Lima, grounding you in the rhythms, rites and culture of the region's pre-Columbian civilisations. The Unesco World Heritage Site is considered by archaeologists to be one of the oldest urban centres in all of the Americas. *www.zonacaral.gob.pe; 3hr 30min by car.*

06 Huacachina
A firmly established backpacker favourite, with an appropriately late-night party scene and international feel, Huacachina is a perfect-looking desert oasis where a dreamy lagoon is ringed by verdant palm trees. By day, intrepid adventurers head out for sandboarding on the nearby dunes, sunset hikes or dune-buggy voyages. Come sunset, it's time to let your hair down. *4hr by car.*

FOODIE PERU

Peru's unique cuisine combines farm-fresh ingredients, indigenous traditions, great seafood and global fusion. In the chic restaurants of Lima and other cities, the long-time favourite, ceviche, takes on new levels of presentation and sophistication, while along the coast you can expect simple ingredients, freshly caught. Another curiosity is cuy (guinea pig), available anywhere but best eaten in the highlands. Beyond that, Peru's link with Asia has ushered in a new school of fusion cooking, with Chinese, Japanese and Thai merging with national traditions that value staple crops such as quinoa and Andean tubers.

DELICIOUS SOUTH AMERICAN FOOD & DRINK ESCAPES

From floating oyster bars and rural wineries to snack stands on the beach and hole-in-the-wall sandwich shops, the opportunities to eat and drink well in South America are endless.

Wine Country, Mendoza, Argentina

A five-course alfresco lunch with wine pairings and a view of the Andes? That's the deal at the rustic but gourmet Bodega La Azul. *Si, por favor. www.bodegalaazul. com; lunch Tue-Sun; approx US$35-40 per person; 1hr 30min from Mendoza by car.*

Floating Oyster Bar, Morro de São Paulo, Brazil

The paradisiacal island of Morro de São Paulo is also the gateway to the Tinharé archipelago. On a speedboat tour, you'll snorkel around a coral reef and have lunch at a floating oyster bar. *2hr from Salvador da Bahia by boat.*

Chivito, Piriápolis, Uruguay

You can get a good *chivito* – the national sandwich of Uruguay, piled high with sliced steak and topped with a fried egg – in the big city. But you'll find even heartier home-made versions in smaller beach towns like Piriápolis. *1hr 30min from Montevideo by car or bus.*

Gaucho-style Barbecue, Argentina

If you want a glimpse of traditional gaucho culture, head to an estancia (ranch) like Los Dos Hermanos, where you'll gallop across the fields on horseback and enjoy a classic asado (barbecue). *www.estancialosdoshermanos. com; 1hr 30min from Buenos Aires by car.*

Seafood, Isla Negra, Chile

The coast around Valparaíso – on the way to Pablo Neruda's house at Isla Negra – is lined with family-run seafood restaurants serving up traditional dishes like *machas a la parmesana* (razor clams baked in parmesan) and freshly grilled *locos* (abalone). *1hr 10min from Valparaíso by car.*

Fresh Fruit & Fire-water, Santa Marta, Colombia

Take a break from city life on the laid-back beaches of Santa Marta, where vendors sell fresh fruit beneath the palm trees – and locals bring their own aguardiente (firewater) to mix it with. *2hr 20min from Barranquilla by car.*

Chocolate & Coffee, Ecuador

Baños is an idyllic destination for several reasons – artisanal Ecuadorian chocolate is just one of them. Try the local speciality of chocolate caliente (hot chocolate) at a cafe and shop like Arome Chocolate & Coffee. *www.arome.com.ec; 3hr from Quito by car.*

Slow Food & Pisco Sours, Valle de Lurín, Peru

Lima is a magnet for foodies, but gourmet pleasures and a slower pace await in the restaurants and bars of the nearby Valle de Lurín, where slow food pairs well with potent pisco sours. *1hr from Lima by car.*

Coffee, Sul de Minas, Brazil

One of the largest coffee-producing regions in Brazil is just north of the capital. Whether you plan a coffee-tasting side trip or are just passing, don't miss an espresso from Sul de Minas. *3hr+ from São Paulo Valle de Lurín by car.*

Chardonnay, Casablanca Valley, Chile

One of Chile's several wine-producing regions is conveniently located just outside the capital city, along the highway to the beach: reserve ahead at one or more of the wineries in Casablanca Valley to sample local chardonnay and sauvignon blanc. *1hr from Santiago by car or bus.*

N

0 — 60 km
0 — 40 miles

04 30 miles
    (50km)

Victoria

Paime

Topaipi

Ubaté

Mariquita   Honda

Chocontá

Guaduas

01

Sesquile

Zipaquira

Guayata

Armero

Cajica

Santa Maria

Alban

Guasca   03

Facatativa

Puli

La Mesa   02

★ BOGOTÁ

20 miles (30km)

40 miles (60km)

60 miles (95km)

Choachi

Medina

Fusagasuga

Girardot   05

Guayabetel

06

Veracruz

El Espinal

Villavicencio

Guamo

San Juan

Pompeya

Cabrera

Acacias

Villarica

Rancho Alegre

Castillo

COLOMBIA

Adventures on horseback, healing waters, an underground cathedral, a towering waterfall... There's so much to see and do around Bogotá that you might want to cut your urban sightseeing short for the beautiful regions that surround the city.

BOGOTÁ

● ARTS & CULTURE ● HISTORY ● OUTDOORS ● FOOD & DRINK ● FESTIVALS & EVENTS ● MUSIC & FILM

── ONE HOUR FROM ──

01 Catedral de Sal, Zipaquirá
A one-of-a-kind landmark, the Salt Cathedral of Zipaquirá is an underground cathedral located inside an old salt mine. Hundreds of feet beneath street level, the interior is carved right into the rock walls, with three main sections that symbolise the birth, life and death of Jesus Christ. The cathedral sits within the larger Parque de Sal complex. *www.catedraldesal.gov.co; 1hr by car or 2hr 30min by bus.*

── TWO HOURS FROM ──

02 La Chorrera
The highest waterfall in Colombia is an easy getaway from Bogotá. You'll need most of the day to actually get to La Chorrera, though: from the entrance of the park, it's a two-hour hike to the waterfall, and another two hours back. Along the trail you'll pass a smaller but equally lovely cascade named El Chiflón, local farmers selling homemade empanadas, and a so-called 'fog forest', blooming with native orchids.
www.lachorrera.com.co; 1hr 30min by car.

03 Estancia San Antonio
A short drive north, a peaceful refuge awaits. The family-run Estancia San Antonio is set in the mountains at an altitude of 2900m. The cooler temperatures are ideal for hiking, leisurely alfresco breakfasts and strolling through the property's orchid gardens – or you could just head to the spa for a massage. There are only five rooms, so book ahead.
www.estanciasanantonio.com; 1hr 30min by car.

04 Hacienda del Salitre
Soak up the healing power of thermal waters in Paipa, a spa town northwest of Bogotá. The region is famous for its hot springs, and a particularly wonderful spot to wallow is the colonial-style Hacienda del Salitre, a centuries-old mansion that Colombia declared a national historic monument. The inn has its own thermal pools, romantically set amid stone fountains and terraces. *www.hotelhaciendaelsalitre. com/en; 2hr 30min by car or 3hr by bus.*

── THREE HOURS FROM ──

05 Girardot
Within a three-hour drive of Bogotá, Girardot sits on the banks of Colombia's most important river, Rio Magdalena. Founded in 1881, the town is a popular day trip or overnight destination for outdoorsy types and those interested in history: you can sunbathe and enjoy river views, or climb aboard an antique train (*el tren de la alegría*, or 'the happy train') for a round trip that includes a segment over the famous Enrique Olaya Herrera bridge. *3hr by car.*

06 Campo Ecologico Gramalote
If you've ever dreamed of riding on horseback through rivers in the Colombian countryside – to clarify, we're talking about deep rivers, where the horse is submerged up to his neck, along with the lower half of your body – here's your chance. At Campo Ecologico Gramalote, you can choose from a variety of guided rides. Just be sure to bring a change of clothes.
www.campoecologicogramalote.com; 3hr by car.

GETTING FROM A TO BOGOTÁ

If you're planning a getaway from Bogotá, you'll need to rent a car or take the bus. Be aware that although driving in the countryside can be a great pleasure, driving into the city will almost certainly be a headache – if you're going to rent a car, consider driving straight to your getaway destination from the airport, where most rental companies have offices. Otherwise, take advantage of the good public bus system operating out of Terminal de Transporte de Bogotá (known simply as 'Salitre'). The bus terminal is colour-coded, so look for yellow if you're heading south, red for north, and blue for getaways to the east and west of Bogotá.

INDEX

September 2019
Published by Lonely Planet Global Limited
CRN 554153
www.lonelyplanet.com
10 9 8 7 6 5 4 3 2 1

Printed in China
ISBN 978 1 78868 931 1
© Lonely Planet 2019
© photographers as indicated 2019

Managing Director, Publishing Piers Pickard
Associate Publisher Robin Barton
Commissioning Editors Dora Ball, Nora Rawn
Art Director Daniel Di Paolo
Layout & Design Daniel Di Paolo, Tina García, Kristina Juodenas
Editor Monica Woods
Picture Research Lauren Marchant
Proofreading Karyn Noble
Print Production Nigel Longuet
Thanks to Nick Mee, Christina Webb

Written by Isabel Albiston, Brett Atkinson, Alexis Averbuck, James Bainbridge, Joe Bindloss, Greg Benchwick, Celeste Brash, Alex Butler, Lucy Corne, Marc di Duca, Janine Eberle, Megan Eaves, Samantha Forge, Bridget Gleeson, Ria de Jong, Carolyn Heller, Michael Kohn, Tatyana Leonov, Alex Leviton, Stephen Lioy, Emily Matchar, Rebecca Milner, Karyn Noble, Trisha Ping, Etain O'Carroll, Helena Smith, Andrea Shulte-Peevers, Ryan Ver Berkmoes.

Lonely Planet Offices

STAY IN TOUCH lonelyplanet.com/contact

Australia
The Malt Store, Level 3,
551 Swanston St, Carlton, Victoria 3053
T: 03 8379 8000

USA
124 Linden St, Oakland,
CA 94607
T: 510 250 6400

Ireland
Digital Depot, Roe Lane (Off Thomas Street)
The Digital Hub,
Dublin 8, D08 TCV4

Europe
240 Blackfriars Rd,
London SE1 8NW
T: 020 3771 5100

MIX
Paper from
responsible sources
FSC™ C021741
www.fsc.org

Paper in this book is certified against the Forest Stewardship Council™ standards. FSC™ promotes environmentally responsible, socially beneficial and economically viable management of the world's forests.